GRAPHING POWER

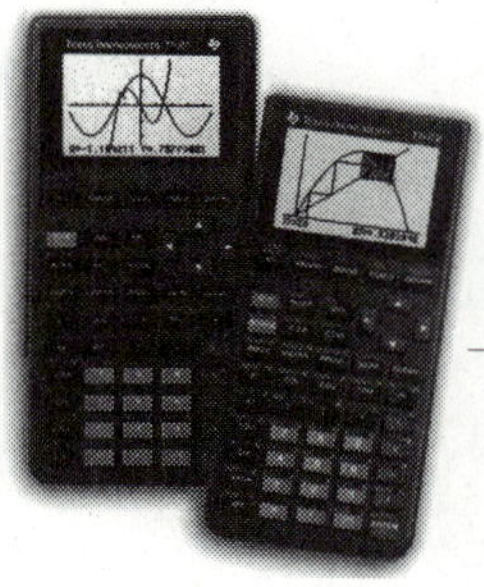

Middle School Activities for the TI-81 and TI-82

The Graphing Technology in Mathematics Project

Dale Seymour Publications

Project Editor: Joan Gideon
Photo TI-81 and TI-82 reprinted courtesy of Texas Instruments, Inc.

Production Coordinator: Claire Flaherty
Text and Cover Design: Lisa Raine
Composition: Andrea Reider
Cover Image: Thomas Lochran, Image Bank

Published by Dale Seymour Publications®, an imprint of the Alternative Publishing Group of Addison-Wesley Publishing Company.

Printed in the United States of America.

Order Number DS25300
ISBN 0-86651-826-6

4 5 6 7 8 9 10–ML–00 99 98 97

DALE SEYMOUR PUBLICATIONS®
P.O. BOX 10888
PALO ALTO, CA 94303

This Book is Printed on Recycled Paper

The Graphing Technology in Mathematics Project

Sponsored by

Trinity University, The San Antonio Independent School District, and The North East Independent School District, San Antonio, Texas

This Project was made possible by a grant from the Dwight D. Eisenhower Mathematics and Science Programs Texas Higher Education Coordinating Board.

Project Director

David D. Molina, Ph.D., Trinity University

Project Co-Directors

Mary Ann Sample, San Antonio Independent School District
Lu Ann Weynand, North East Independent School District

Graduate Assistants

Wallace Ross, Trinity University
Stacy Dill, Trinity University
Karen Stenson, Trinity University

Editor

Alma Aguirre, Thomas Jefferson High School

Contributing Editor

Laurie Bergner, Robert E. Lee High School

Participants and Contributing Writers

Shelia Lloyd
Brackenridge HS

Linda Patterson
Brackenridge HS

Janet Starkey
Brackenridge HS

Sylvia Carreón
Burbank HS

Ascencion Gonzalez
Burbank HS

Yolanda Hughes
Burbank HS

Michael Russell
Burbank HS

Flor de Maria Wukovits
Burbank HS

Roland Rios
Churchill HS

Nancy Cogdell
Connell MS

Debbie Fisher
Connell MS

Susan Tustison
Connell MS

Evelyn Williams
Cooper MS

Diana Harwood
Driscoll MS

Mike Hansen
Eisenhower MS

Vera Hayes
Eisenhower MS

Maria E. Cantú
Edison HS

Magda Gonzales
Edison HS

Marie Graham
Edison HS

Susan Ng
Edison HS

Diana Hughes
Garner MS

Rebecca Moch
Harris MS

Kim Rodriguez
Harris MS

George Watkins
Healy Murphy

Gary Frank Clark
Highlands HS

Shirley Rich
Highlands HS

Naida Brueland
Irving MS

Anna Pilgrim
Irving MS

Karen Allbright
Jackson MS

Lee Anne Leister
Jackson MS

Alma Aguirre
Jefferson HS

Cathy Morgan
Judson HS

Annette Plowman
Judson HS

Rosie Sabala
King MS

Ana Thompson
Krueger MS

Valorie Fayfich
Lanier HS

Carole Gould
Lee HS

Shan Schanding
Lee HS

Paul Tisdel
Lee HS

Anthony Plummer
Lee HS

Juanita B. Martinez
Lowell MS

Maria T. Muñoz
Lowell MS

Michael Maltrud
Marshall HS

Andy Davis
Navarro AC

David Monteith
Nimitz MS

Joe Rosseett
Page MS

June Draper
Rogers MS

Sergio Perales
Roosevelt HS

William Lauderdale
Sam Houston HS

James F. (Skip) Lee
Sam Houston HS

Karla Escalera
St. Gregory

Stella Vasquez
St. Gregory

Patricia Gomez
Tafolla MS

Betsy Burke
Trinity Univ.

Erin Roberson
Trinity Univ.

Shannon Walsh
Trinity Univ.

Leslie Cate
Twain MS

Rosemary Hidalgo
Twain MS

Andrew Stepanski
Twain MS

Rudy Tamez
Twain MS

Evangeline Tovar
Wheatley MS

Sandy Rothrock
White MS

Mary Lynne Montgomery
White MS

Peter Barrera
Whittier MS

Preface

The graphing calculator gives students the graphing power to explore mathematics concepts. With it students can see the slope of a line change as they key in different coefficients of x; they can watch the phase shift and the amplitude vary as they alter the coefficients in trigonometric functions. By enabling students to quickly generate examples, the graphing calculator allows them to see patterns, discover relationships, and validate conjectures. Educators are advocating use of this powerful tool in mathematics classrooms.

For two summers, groups of teachers gathered at Trinity University in San Antonio, Texas, to study the use of the graphing calculator in the middle school and high school. Participants generated lessons they could use in their classrooms. The two books in the *Graphing Power* series represent the results of their efforts. The work done by these teachers will help you make effective use of graphing technology in your classroom. These activities are suggestions and models for lessons that make the best use of the calculator; they will give you ideas for developing your own lessons. Use them in conjunction with concrete materials, other technologies, and traditional lessons.

These activities are intended for use in a classroom where students are encouraged to explore and discover, preferably in groups of two to four. Instructions in the lessons provide help using the TI-81 and TI-82, but some experience with the graphing calculator is assumed. Have calculator manuals available for reference as you work through the lesson before presenting it to your class and as your students complete the activity.

If you are working with gifted students or teaching pre-algebra or algebra, you can find additional activities in *Graphing Power: High School Activities for the TI-81 and TI-82*.

Contents

TOPIC 1

Working with Numbers

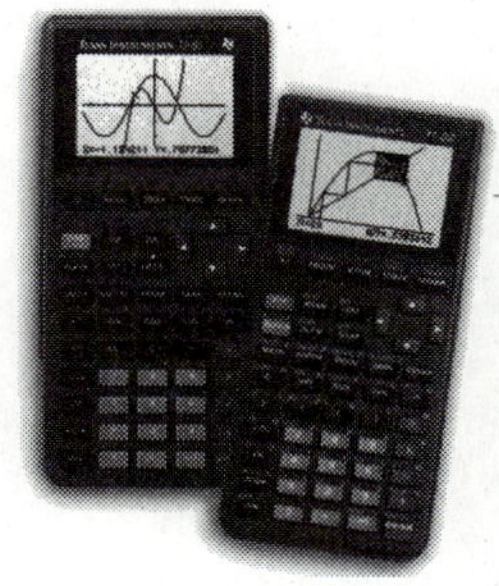

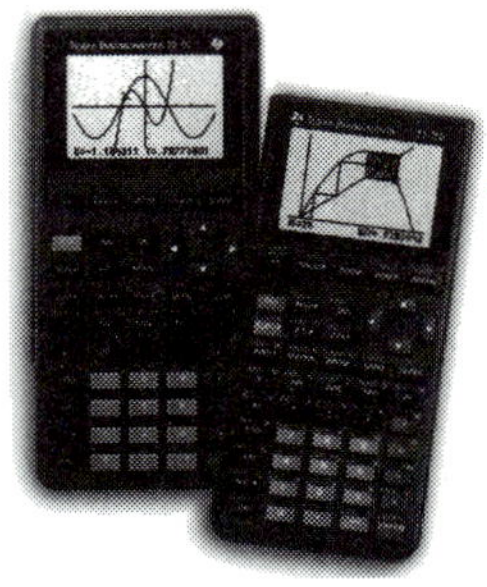

Multiplying by Powers of 10

Using "Powers of 10" students will

- Use the exponent features on the calculator to explore the results of multiplying by powers of 10
- Describe the effects of multiplying by powers of 10 with both positive and negative exponents
- Rewrite a number in standard form to a product using a power of 10
- Develop a set of rules for multiplying by powers of 10
- Describe their observations and results in writing

Teaching the Lesson

- Set the decimal point to floating in the [MODE] menu.
- Note the difference between the [(–)] key and the [–] key when calculating negative exponents.
- The focus of this lesson is on preparing students for scientific notation, it is not necessarily on proper scientific notation.

Solutions

9. Mars—approximately 730 days
   Pluto—approximately 55,600 days
10. 1.24^{-3}

Powers of 10

1. Enter the following.

a. 5.7×10^2, input 5.7 [X] 10 [^] 2, press [ENTER]

b. 6.4×10^{-3}, input 6.4 [X] 10 [^] [(–)] 3, press [ENTER]

2. Enter the following and observe the changes in the product.

a. 3.5×10^1 b. 3.5×10^2

c. 3.5×10^3 d. 3.5×10^4

3. Answer the following questions.

a. What is common about the products?

b. How do the products differ?

c. What do you predict these products to be?

3.5×10^5 3.5×10^9

d. What is the pattern for multiplying by a positive power of 10?

4. Enter the following and observe the changes in the product.

a. 62×10^0 b. 62×10^{-1}

c. 62×10^{-2} d. 62×10^{-3}

5. Answer the following questions.

a. What is common about the products?

b. How do the products differ?

c. What do you predict the products of the following to be?

62×10^{-4} 62×10^{-7}

d. What is the pattern for multiplying by a negative power of 10?

6. Enter the following and observe the changes in the product.

a. 35×10^0 b. 2.8×10^0

c. 17×10^0 d. 125×10^0

7. Answer the following questions.

a. What is common about the products?

b. How do the products differ?

c. What is the value of 10^0?

d. What is the pattern for multiplying by 10^0?

8. Write the following numbers using a powers of 10.

a. 12400 b. 320000 c. 0.0045

d. 0.00003 e. 12500000 f. 12

g. 12000000 h. 0.000000032 i. 0.34

Verify your answers by entering them into the calculator.

9. It takes approximately 5 days for a spaceship to travel from the earth to the moon, a distance of 2.4×10^5 miles. At this rate, about how many days would it take to travel from earth to Mars, a distance of 3.5×10^7 miles? To Pluto, a distance of 2.67×10^9?

10. Use a power of 10, one negative sign, a decimal point, and the digits 1, 2, 3 and 4 to write the number that is closest to 0.001.

Scientific Notation

Using "Scientific Notation" students will

- Change numbers from standard form to scientific notation
- Change numbers from scientific notation to standard form
- Check solutions using the calculator
- Recognize the scientific notion format shown on the viewing screen
- Convert the calculator's scientific notation format to common scientific notation form
- Convert the calculator's scientific notation format to standard form
- Use the scientific mode capabilities on the calculator to convert numbers in standard form to scientific form on the calculator

Teaching the Lesson

- Use [MODE] to change the calculator into scientific notation format. Remind students to set the calculator back to Norm at the end of the lesson.
- The calculator's scientific notation format will display an "E" followed by a number indicating the exponent value for scientific notation form.

Scientific Notation

1. Compute the following using your graphing calculator.

a. 4.58×10^4 b. 1.25×10^3 c. 7×10^5

2. How would you enter 4.8×10^{-3} on the graphing calculator?

3. Compute the following.

a. 3.4×10^{-2} b. 1.8×10^{-3}

4. To change from standard notation to scientific notation.

Press [MODE]

Move your cursor to SCI

Press [ENTER]

Press [CLEAR]

You are now ready to enter numbers in scientific notation. Input each number followed by a [ENTER]. Record the result.

a. 4.58×10^4 b. 1.25×10^3 c. 7×10^5 d. 4.8×10^{-3}

________ ________ ________ ________

5. What similarities do your results have?

6. What do you think the "E" in your display represents?

7. What do the numbers after the "E" represent?

8. Enter these numbers, and record the calculator's display.

a. 32060 b. 12400 c. 425 d. 12

________ ________ ________ ________

9. What similarities do your results have?

10. Convert the calculator display to scientific notation for each number in question 8.

a. ________ b. ________ c. ________ d. ________

11. Convert the calculator display to scientific notation for these numbers.

a. 0.478 b. 0.045 c. 0.0000032

________ ________ ________

12. What similarities do your results have?

13. Compare your answers in question 4 to your answers in question 8. What conclusions can you draw?

14. To see if your conclusion is correct, convert the following to scientific notation without the calculator, then use the calculator to verify your results.

	You	Calculator
a. 3200	________	________
b. 13900	________	________
c. 0.0307	________	________
d. 0.02004	________	________

Order of Operations I

Using "Order I" students will

- Use the edit commands to explore the effects of parentheses on numerical expressions
- Develop guidelines for following order of operations
- Use order of operations to solve for a given solution

Teaching the Lesson

- Remind students to use the edit commands to save time.
- The students may need some guidance to see which operations were performed first. Devote time to ensuring that the students understand the order of the operations performed.

Order I

1. Evaluate the following expressions on your calculator and record your answers. Write down which operation (+, –, ×, ÷) was performed first and whether or not it was in parenthesis.

	Answer	First Operation Occurring	Was the First Operation in ( )?
$2 \times 6 + 1$			
$(2 \times 6) + 1$			
$2 \times (6 + 1)$			
$14 \div 2 + 5$			
$(14 \div 2) + 5$			
$14 \div (2 + 5)$			
$9 - 3 \div 3$			
$(9 - 3) \div 3$			
$9 - (3 \div 3)$			
$24 \div 8 - 2$			
$(24 \div 8) - 2$			
$24 \div (8 - 2)$			
$12 + 9 \div 3$			

	Answer	First Operation Occurring	Was the First Operation in ( )?
$(12 + 9) \div 3$			
$12 + (9 \div 3)$			
$15 + 5 \times 2$			
$(15 + 5) \times 2$			
$15 + (5 \times 2)$			
$6 \times 2 \div 2$			
$(6 \times 2) \div 2$			
$6 \times (2 \div 2)$			
$5 + 7 \times 4 \div 2$			
$(5 + 7) \times 4 \div 2$			
$5 + (7 \times 4) \div 2$			
$(5 + 7) \times (4 \div 2)$			
$5 + 7 \times (4 \div 2)$			

2. Of the four operations—addition, subtraction, multiplication, division—in which order do you think they are performed?

First ____________________ Second ____________________

Third ____________________ Fourth ____________________

3. Which operations in parentheses did not change the solution of the expression?

4. Which operations in parentheses did change the solution of the expression?

5. What effect did the parentheses have on the order of operations?

6. What effect do the parentheses have on the order in which the operations are performed? What observations have you made to support your answer?

7. When the calculator is evaluating these expressions, which operation is performed first?

$3 + 4 \times 2$ ________ $3 \times 4 + 2$ ________

8. When the calculator is evaluating these expressions, which operation is performed first?

$12 + 4 \div 2$ ________ $12 \div 4 + 2$ ________

9. When the calculator is evaluating these expressions, which operation is performed first?

$4 + 3 - 2$ ________ $4 - 3 + 2$ ________

10. When the calculator is evaluating these expressions, which part is performed first?

$(4 + 3) - 2$ ________ $4 + (8 - 6)$ ________

11. When adding and subtracting, does order matter? Which occurs first? State an example.

12. When multiplying and dividing, does order matter? Which occurs first? State an example.

13. When adding and multiplying, does order matter? Which occurs first? State an example.

14. When subtracting and dividing, does order matter? Which occurs first? State an example.

15. Considering the four operations—addition, subtraction, multiplication, division—and the use of parentheses, what order should you follow to evaluate an expression?

First ______________________________

Second ______________________________

Third ______________________________

16. Classify each statement as True or False. If False, make the statement true by inserting parentheses where necessary.

a. $5 \times 4 + 6 = 50$

b. $24 \div 3 + 5 \times 2 = 6$

c. $4 \times 6 + 3 \times 3 = 33$

d. $6 \times 12 - 5 \times 8 = 32$

e. $3 \times 4 - 2 \times 3 = 18$

f. $7 + 6 \times 9 - 5 = 31$

Order of Operations II

Using "Order II" students will

- Use the edit commands to explore expressions using the order of operations
- Explore expressions using order of operations
- Use the order of operations to solve for a given solution
- Use guess and check to solve questions

Teaching the Lesson

- Remind students to use the edit commands to save time. Encourage them to find patterns and to think about possible solutions, rather than to simply push buttons on the calculator.
- The directions for numbers 10–15 allow the use of decimal points, square roots, and exponents.

Solutions (Answers will vary.)

$4 + 4 \div 4 - 4 = 1$

$4 \div 4 + 4 \div 4 = 2$

$(4 + 4 + 4) \div 4 = 3$

$(4 - 4) \div 4 + 4 = 4$

$(4 \times 4 + 4) \div 4 = 5$

$(4 + 4) \div 4 + 4 = 6$

$4 - 4 \div 4 + 4 = 7$

$4 + 4 + 4 - 4 = 8$

$4 + 4 + 4 \div 4 = 9$

$4 + 4 + 4 \div \sqrt{4} = 10$

$4^2 - 4 - (4 \div 4) = 11$

$4^2 - 4 \times (4 \div 4) = 12$

$4^2 - 4 + (4 \div 4) = 13$

$4 + 4 + 4 + \sqrt{4} = 14$

$4^2 - 4 \div \sqrt{4} \div \sqrt{4} = 15$

$4 - 4 + 4 \times 4 = 16$

$4 \times 4 + 4 \div 4 = 17$

$4^3 \div 4 + 4 \div \sqrt{4} = 18$

$4^2 + 4 - 4 \div 4 = 19$

$4 + 4^2 + 4 - 4 = 20$

$4^2 + 4 + 4 \div 4 = 21$

$4 + 4 + 4 \times 4 = 24$

$4 + 4^2 + 4 + 4 = 28$

$4^2 + \sqrt{4} + 4^2 - 4 = 30$

$4^2 \times \sqrt{4} - 4 \div 4 = 31$

$4^2 + 4 + 4^2 - 4 = 32$

Order II

Use your calculator to find the solution.

Fill in the blanks with the correct symbol to make each statement true. You may use the +, –, ×, ÷, and parentheses between the 4s to obtain the given solution.

1.	4	______	4	______	4	______	4	=	1
2.	4	______	4	______	4	______	4	=	2
3.	4	______	4	______	4	______	4	=	3
4.	4	______	4	______	4	______	4	=	4
5.	4	______	4	______	4	______	4	=	5
6.	4	______	4	______	4	______	4	=	6
7.	4	______	4	______	4	______	4	=	7
8.	4	______	4	______	4	______	4	=	8
9.	4	______	4	______	4	______	4	=	9

Fill in the blanks with the correct symbol to make each statement true. For some of the following, you may use a decimal point, square roots, and exponents, as well as the +, –, ×, ÷, and parentheses between the 4s to obtain the given solution.

10.	4	______	4	______	4	______	4	=	10
11.	4	______	4	______	4	______	4	=	11
12.	4	______	4	______	4	______	4	=	12
13.	4	______	4	______	4	______	4	=	13
14.	4	______	4	______	4	______	4	=	14
15.	4	______	4	______	4	______	4	=	15

16. Continue with the numbers 16–32 on the back; some numbers may not have a solution.

17. Which solutions were the easiest to find? Why?

18. Find "4 expressions" for numbers greater than 32.

Fractions and Decimals

Using "Where Do I Belong?" students will

- Graph fractions on a number line
- Graph decimals on a number line
- Graph decimals and fractions together on a number line
- Use the Edit and Xsort function to order decimals
- Relate decimals to their fraction equivalent and order the fractions
- Construct a number line using fractions and decimal values from the chart

Teaching the Lesson

- Questions 1–9 serve as a review and can be done with or without a calculator.
- The keystrokes for SORTing are very different for the TI-81 and the TI-82. Both are included in the lesson.
- Each group will need a long piece of adding machine tape, butcher paper, or chalkboard space for their number line.

Where Do I Belong?

1. Divide the following number line into halves. Label $\frac{1}{2}$, $1\frac{1}{2}$, $2\frac{1}{2}$.

2. Now label the following number line with thirds. Remember "thirds" means three parts not three lines.

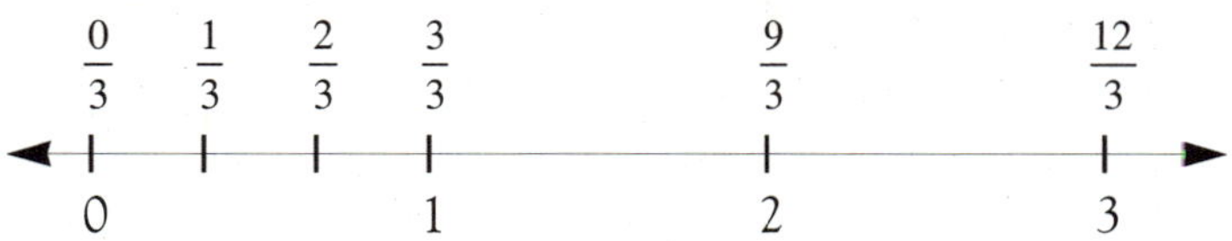

3. Graph fourths on this number line.

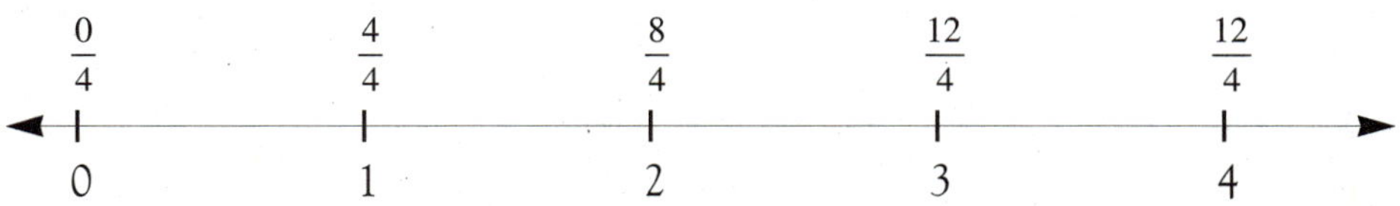

4. Fill in the blanks

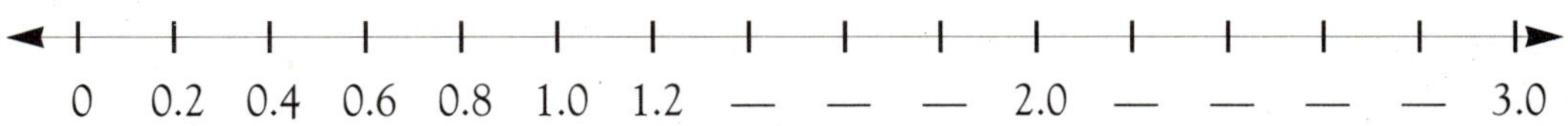

5. Describe the pattern.

6. Mark the midpoint between 0.4 and 0.6, what decimal does it represent? Why?

7. Break a dollar into 10 equal parts. How much is $\frac{1}{10}$ of a dollar?

8. Graph \$1.00 by tenths on a number line numbered from \$0.00 to \$1.00.

9. Graph the following on the number line below.

$1.50 $1.70 $1.30 $0.50 $0.35 $2.75 $2.60 $2.20 $0.05 $0.90

To find the decimal equivalent, use the calculator to divide the numerator by the denominator.

10. Label the fractions with their decimal equivalents.

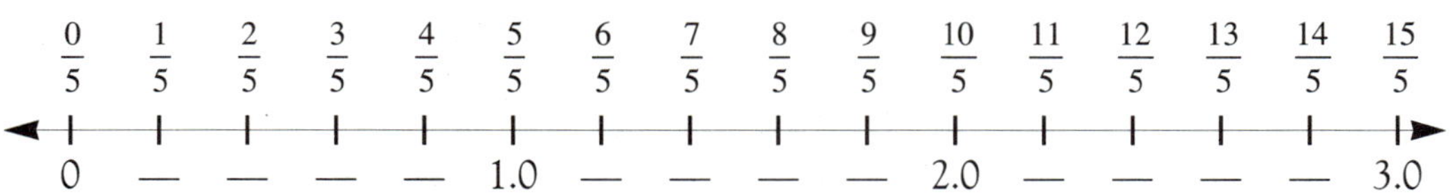

11. Compare the fractions above. What do you notice?

12. Compare the decimal values above. What do you notice?

13. Fill in the missing fractions. Label the fractions with their decimal equivalents.

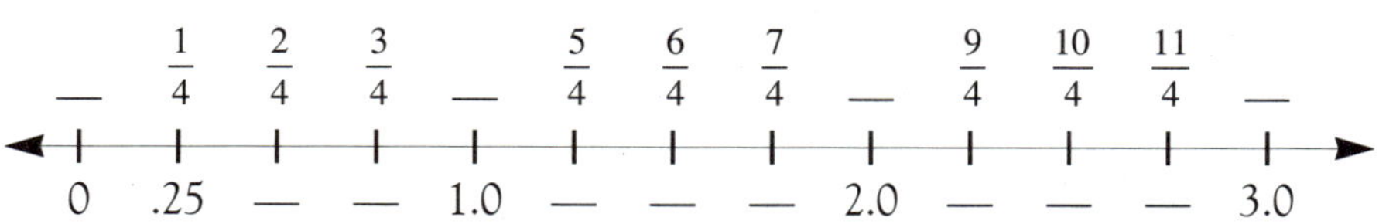

14. Compare the fractions above. What do you notice?

15. Compare the decimal values above. What do you notice?

16. Compare the fraction and decimal values and describe the pattern you observe.

17. Create your own number line for the numbers 0–3, indicating each $\frac{1}{2}$. Predict a pattern for each fraction and decimal. Verify your prediction using the calculator.

18. Order the decimals below from least to greatest. Record them in the table under Decimal Prediction.

3.5	0.25	2.333	0.5	0.667	1.5	3.667	.0333
0.75	0.8	0.4	0.2	0.6	2.75	1.8	1.33

A Decimal Prediction	B. Decimal Sort	C. Fraction Prediction	D. Fraction Sort

19. Use the STAT menu of your calculator to enter the above data. Select DATA and press ENTER to edit the data. Enter the decimals. After each entry, press ENTER twice if using a TI-81.

If you're using a TI-82, use the STAT menu and edit your data by placing the decimals in L_1.

20. If using a TI-81: To order the decimals using the calculator, access the [STAT] menu, select DATA, and choose xSort, then [ENTER]. This will sort the decimals in order from smallest to largest. To see the sorted data, access the [STAT] menu again, select DATA, and choose Edit. As you scroll down, you can view the decimals.

If using a TI-82: To order the decimals using the calculator, use the [STAT] menu and EDIT data by using SORT A. Sort L_1. This will sort the decimals in order from smallest to largest. To see the sorted data, access the [STAT] menu and select EDIT.

21. Record this order on the chart. How good was your prediction? Can you explain any anomalies?

22. Match each of the following fractions to a decimal on the chart (use the sorted column). Predict each fraction and decimal equivalent. How are you making your prediction?

$\frac{3}{2}$ $\frac{1}{3}$ $\frac{9}{5}$ $\frac{3}{4}$ $\frac{2}{5}$ $\frac{2}{3}$ $\frac{11}{4}$ $\frac{4}{5}$

$\frac{1}{4}$ $\frac{11}{3}$ $\frac{1}{5}$ $\frac{7}{2}$ $\frac{1}{2}$ $\frac{3}{5}$ $\frac{4}{3}$ $\frac{7}{3}$

23. Use your calculator to convert the fractions to decimals and complete the chart. How good was your prediction? Explain?

24. What did you notice about $\frac{1}{3}$, $\frac{2}{3}$, and $\frac{11}{3}$? What is this type of decimal called? Why?

25. Draw a large number line for the classroom display. Graph the decimals and fractions from the chart on this number line.

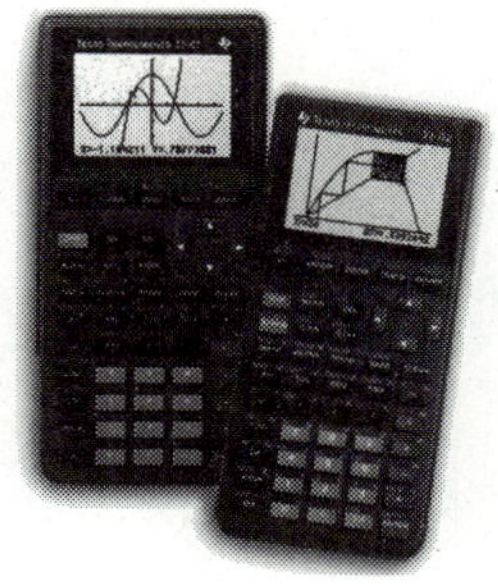

Decimals and Percents

Using "Choose a Bank" students will

- Use given information from various banks to decide which offers the best service for the best price
- Use the edit commands to calculate bank charges
- Support reasons for choosing a given bank

Teaching the Lesson

- Students will need to review the following vocabulary.

  Service Charge: A monthly fee assessed on an account.

  Cost per Check (per item charge): The cost for each check drawn on an account that clears the bank.

  Minimum Balance: The minimum amount required to maintain an account without a service charge. A service charge is generally assessed when a balance falls below the minimum balance. Sometimes a minimum balance is required to earn interest.

Choose a Bank

1. Complete the chart using the graphing calculator's editing capabilities. Decide which bank offers the best service for the best price.

First Bank

No interest is paid on the balance.
There are no service charges.
The cost per check is \$0.25.

Second Bank

Interest is paid on the beginning balance at a rate of 3% per year paid monthly.
($\frac{1}{4}$% per month = 0.0025)

The service charge is \$5.00 per month.
The cost per check is \$0.25.
No minimum balance required.

Third Bank

Interest is paid on the beginning balance at a rate of 4.5% per year paid monthly.
($\frac{3}{8}$% per month = 0.00375)

The are no service charges.
The cost per check is \$0.25. However, if the beginning balance falls below \$500.00, the cost per check charge is \$0.50.

	First Bank				Second Bank				Third Bank			
Beginning Balance	\$650	\$625	\$450	\$150	\$650	\$625	\$450	\$150	\$650	\$625	\$450	\$150
Number of Checks	18	26	20	31	18	26	20	31	18	26	20	31
Interest Earned												
Service Charge												
Total Check Charges												
New Balance												

2. Which bank would you choose for your banking needs?

3. When selecting a bank, there is no right or wrong choice. The choice depends on an individual's banking needs. Why did you choose this bank? Explain your reasons in several paragraphs. Be sure to compare and contrast the banks in your explanation.

4. Were there questions you had about the banks that were not answered? Could you have expanded the chart to help answer them?

5. Go to local banks and collect information about checking account services. Do this same investigation with data from real banks in your area. Which bank would you choose? Why?

Multiplying Terms with Like Bases

Using "Multiplying Powers" students will

- Use the exponent commands to multiply terms with like bases
- Predict which expressions are equivalent
- Test their predictions with the calculator
- Based on observations, develop a rule for multiplying with like bases

Teaching the Lesson

- This lesson can be done in two parts. Part I deals with positive exponents. Part II deals with negative exponents. While each lesson can be used independently, there are references in Part II to the work done in Part I.

Multiplying Powers

Part I

Predict which of the following expressions are equivalent. Circle your prediction from column a, b, or c. Then use your calculator to verify your prediction.

			a	b	c
1.	$3^2 \times 3^4$	=	3^8	3^2	3^6
2.	$6^0 \times 6^2$	=	6^0	6^2	6^6
3.	$5^7 \times 5^3$	=	5^4	5^{10}	5^{21}
4.	$4^8 \times 4^7$	=	4^{15}	4^1	4^{56}
5.	$2^3 \times 2^{10}$	=	2^{30}	2^7	2^{13}
6.	$1^2 \times 1^{13}$	=	1^{11}	1^{26}	1^{15}

Record the pairs of equivalent expressions.

$3^2 \times 3^4 =$ __________ $4^8 \times 4^7 =$ __________

$6^0 \times 6^2 =$ __________ $2^3 \times 2^{10} =$ __________

$5^7 \times 5^3 =$ __________ $1^2 \times 1^{13} =$ __________

7. How are the exponents of the first expressions related to the exponents of the second expressions? Write a rule that expresses this relationship?

8. Use your rule on the following problems. Fill in the proper exponent.

a. $2^5 \times 2^6 = 2^{\square}$ b. $7^2 \times 7^5 = 7^{\square}$ c $4^3 \times 4^2 = 4^{\square}$

Verify your answers with the calculator.

Part II

Predict which of the following expressions are equivalent. Circle your prediction from column a, b, or c. Then use your calculator to verify your prediction.

			a	b	c
1.	$2^3 \times 2^3$	=	2^{18}	2^9	2^2
2.	$2^6 \times 2^{-3}$	=	2^{-18}	2^3	2^{-2}
3.	$2^{-6} \times 2^{-3}$	=	2^{-9}	2^2	2^{18}
4.	$6^{-5} \times 6^2$	=	6^3	6^{-10}	6^{-3}
5.	$5^{-4} \times 5^8$	=	5^{-32}	5^{-2}	5^4
6.	$7^{-3} \times 7^{-7}$	=	7^{-10}	7^{10}	7^4
7.	$4^5 \times 4^{-8}$	=	4^{-40}	4^{-3}	4^{-13}
8.	$3^{-4} \times 3^4$	=	3^{-8}	3^{-16}	3^0

Record the pairs of equivalent expressions.

$2^6 \times 2^3 =$ ____________ $5^{-4} \times 5^8 =$ ____________

$2^6 \times 2^{-3} =$ ____________ $7^{-3} \times 7^{-7} =$ ____________

$2^{-6} \times 2^{-3} =$ ____________ $4^4 \times 4^{-8} =$ ____________

$6^{-5} \times 6^2 =$ ____________ $3^{-4} \times 3^4 =$ ____________

9. Can the rule you developed using positive exponents be applied to negative exponents?

10. Use your rule on the following problems. Fill in the proper exponent.

a. $3^{-7} \times 3^{11} = 3^{\square}$ b. $2^2 \times 2^{-10} = 2^{\square}$ c. $5^{-6} \times 5^{-4} = 5^{\square}$

d. $7^8 \times 7^{-8} = 7^{\square}$ e. $6^{-5} \times 6 = 6^{\square}$ f. $3^4 \times 3^{-7} \times 3^{-2} = 3^{\square}$

Verify your answers with the calculator.

Dividing Terms with Like Bases

Using "Dividing Bases" students will

- Predict which expressions are equivalent
- Use the exponent commands to divide terms by like terms
- Test their predictions with the calculator
- Based on observations, develop a rule for dividing terms with like bases

Teaching the Lesson

- This lesson can be broken down into three parts. The breaks occur after positive exponents, exponents of zero and one, and again when working with negative exponents. There are, however, references to work done in previous sections. Experience working with negative exponents will be helpful when completing Part III.
- The keystrokes can create problems if the students are not careful. There is extensive use of exponents and the [(–)] sign.

Dividing Powers

Part I

Predict which of the following expressions are equivalent. Circle your prediction from column a, b, or c. Then use your calculator to verify your prediction.

			a	b	c
1.	$5^3 \div 5^2$	=	5^6	5^1	5^5
2.	$7^8 \div 7^8$	=	7^{16}	7^{64}	7^0
3.	$4^5 \div 4^3$	=	4^2	4^{15}	4^8
4.	$2^8 \div 2^7$	=	2^{15}	2^1	2^{56}
5.	$3^2 \div 3^1$	=	3^3	3^2	3^1
6.	$6^5 \div 6^3$	=	6^5	6^2	6^{15}
7.	$7^9 \div 7^6$	=	7^{15}	7^{54}	7^3
8.	$10^2 \div 10^2$	=	10^0	10^6	10^4

Record the pairs of equivalent expressions.

$5^3 \div 5^2 =$ __________ $3^2 \div 3^1 =$ __________

$7^8 \div 7^8 =$ __________ $6^5 \div 6^3 =$ __________

$4^5 \div 4^3 =$ __________ $7^9 \div 7^6 =$ __________

$2^8 \div 2^7 =$ __________ $10^2 \div 10^2 =$ __________

9. How are the exponents of the first expressions related to the exponents of the second expressions? Write a rule that expresses this relationship.

10. Use your rule on the following problems. Fill in the proper exponent.

a. $5^4 \div 5^3 = 5^{\square}$ b. $12^7 \div 12^4 = 12^{\square}$ c. $4^3 \div 4^2 = 4^{\square}$

d. $4^3 \div 4^3 = 4^{\square}$ e. $8^5 \div 8^4 = 8^{\square}$

Verify your answers with the calculator.

Part II

1. Look back at your answers to the following problems and fill in the correct exponent.

$7^8 \div 7^8 = 7^{\square}$ $10^2 \div 10^2 = 10^{\square}$ $4^2 \div 4^2 = 4^{\square}$

2. What was the exponent in each of these solutions? What is the value of each solution?

3. Evaluate each of the following expressions with your calculator

$3^0 =$ _______ $8^0 =$ _______ $6^0 =$ _______ $15^0 =$ _______ $0^0 =$ _______

4. Looking at the pattern, what rule can you write about the value of any number raised to the zero power?

5. Look back at your answers to complete these problems and fill in the correct exponent.

$5^3 \div 5^2 = 5^{\square}$ $3^2 \div 3^1 = 3^{\square}$ $5^4 \div 5^3 = 5^{\square}$

6. What was the exponent in each of these solutions? What is the value of each solution?

7. Evaluate each of the following expressions with your calculator.

$1^1 =$ _______ $7^1 =$ _______ $8^1 =$ _______ $17^1 =$ _______ $0^1 =$ _______

8. Looking at the pattern, what rule can you write about the value of any number raised to the first power?

Part III

Predict which of the following expressions are equivalent. Circle your prediction from column a, b, or c. Then use your calculator to verify your prediction.

			a	b	c
1.	$7^6 \div 7^3$	=	7^9	7^3	7^2
2.	$2^{-2} \div 2^4$	=	2^4	2^6	2^{-6}
3.	$5^3 \div 5^6$	=	5^{-3}	5^{-9}	5^3
4.	$6^{10} \div 6^{-7}$	=	6^{-4}	6^{-3}	6^{-2}
5.	$3^{-4} \div 3^{-6}$	=	3^{-12}	3^2	3^{-2}
6.	$2^{-8} \div 2^{-2}$	=	2^4	2^6	2^{-6}
7.	$10^4 \div 10^{-2}$	=	10^2	10^{-8}	10^6
8.	$9^{-7} \div 9^{-7}$	=	9^0	9^7	9^{-14}

9. Record the pairs of equivalent expressions.

$7^6 \div 7^3 =$ ____________ $3^{-4} \div 3^{-6} =$ ____________

$2^{-2} \div 2^4 =$ ____________ $2^{-8} \div 2^{-2} =$ ____________

$5^3 \div 5^6 =$ ____________ $10^4 \div 10^{-2} =$ ____________

$6^{10} \div 6^{-7} =$ ____________ $9^{-7} \div 9^{-7} =$ ____________

10. Does the rule you developed for positive exponents work with negative exponents?

11. Use your rule on the following problems. Fill in the proper exponent.

a. $6^5 \div 6^7 = 5^{\square}$ b. $7^8 \div 7^{12} = 7^{\square}$ c. $4^{-3} \div 4^2 = 4^{\square}$

e. $3^{-3} \div 3^{-2} = 3^{\square}$ f. $5^5 \div 5^{-3} = 5^{\square}$ g. $9^{-2} \div 9^{-6} = 9^{\square}$

h. $2^{-6} \div 2^4 = 2^{\square}$ i. $3^{-4} \div 3^4 = 3^{\square}$ j. $5^{-3} \div 5^{-3} = 5^{\square}$

Verify your answers with the calculator.

Raising Powers to Powers

Using "The Power of a Power" students will

- Predict which expressions are equivalent
- Test their predictions with the calculator
- Based on observations, develop a rule for raising powers to powers

Teaching the Lesson

- This lesson can be done in two parts. There is, however, a reference in Part II to work done in Part I. This lesson makes extensive use of keystrokes for exponents and parentheses. Remind students that there is a difference between the [(–)], negative key, and the [–], subtract key.

The Power of a Power

Part I

Predict which of the following expressions are equivalent. Circle your prediction from column a, b, or c. Then use your calculator to verify your prediction.

			a	b	c
1.	$(2^2)^3$	=	2^5	2^6	2^8
2.	$(5^4)^2$	=	5^6	5^{16}	5^8
3.	$(7^1)^4$	=	7^4	7	7^8
4.	$(12^3)^0$	=	12^0	12^3	12^1
5.	$(4^6)^2$	=	4^8	4^4	4^{12}
6.	$(8^5)^1$	=	8^6	8^5	8^4

Record the pairs of equivalent expressions.

$(2^2)^3 =$ __________ $(12^3)^0 =$ __________

$(5^4)^3 =$ __________ $(4^6)^2 =$ __________

$(7^1)^4 =$ __________ $(8^5)^1 =$ __________

7. How are the exponents of the first expressions related to the exponents of the second expressions? Write a rule that expresses this relationship.

8. Use your rule on the following problems. Fill in the proper exponent.

a. $(3^2)^4 = 3^{\square}$ b. $(9^3)^3 = 9^{\square}$ c. $(2^6)^5 = 2^{\square}$

Verify your answers with the calculator.

Part II

Predict which of the following expressions are equivalent. Circle your prediction from column a, b, or c. Then use your calculator to verify your prediction.

			a	b	c
1.	$(3^{-2})^3$	=	3^{-6}	3	3^{-8}
2.	$(2^4)^{-2}$	=	2^{-2}	2^2	2^{-8}
3.	$(4^{-4})^{-2}$	=	4^{-6}	4^8	4^{16}
4.	$(5^{-3})^2$	=	5^{-1}	5^9	5^{-6}
5.	$(3^3)^{-5}$	=	3^{-15}	3^{-2}	3^8
6.	$(2^{-3})^{-3}$	=	2^{-6}	2^9	2^{-27}

Record the pairs of equivalent expressions.

$(3^{-2})^3 =$ __________ $(5^{-3})^2 =$ __________

$(2^4)^{-2} =$ __________ $(3^3)^{-5} =$ __________

$(4^{-4})^{-2} =$ __________ $(2^{-3})^{-3} =$ __________

7. Is the rule that you developed in Part I still true? Does working with negative exponents change the rule?

8. Use your rule on the following problems. Fill in the proper exponent.

a. $(3^{-5})^2 = 3^{\square}$ b. $(2^3)^{-4} = 2^{\square}$ c. $(5^{-2})^{-3} = 5^{\square}$

Verify your answers with the calculator.

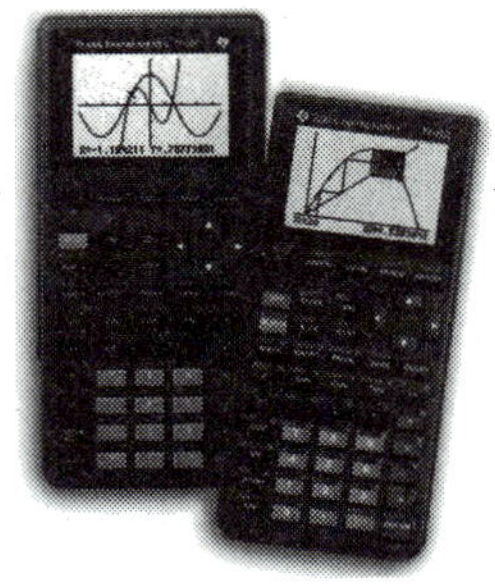

Evaluating Expressions

Using "Express Yourself" students will

- Explore the relationship between pairs of numbers in a table
- Explore patterns in tables
- Describe in words the pattern for each table
- Solve a real life problem by developing a table, exploring a relationship, describing the relationship, and applying it
- Use the calculator to explore expressions
- Write expressions based on patterns in the tables

Teaching the Lesson

- Students may need help in storing values for x when evaluating the expressions.

Express Yourself

Explore with your calculator the relationship between the numbers in the two columns. Explain this relationship in complete sentences.

1.

Pies	Apples
1	6
2	12
3	18
4	24
5	30
6	36

2.

Dogs	Fleas
0	0
1	250
3	750

3.

x	Answer
3	7
2	6
1	5
0	4
–1	3

4.

Amount Paid	Change
100	75
90	65
80	55
70	45
60	35

Discover the pattern of each table and then complete the table.

5.

Movie Goer	Ticket Price
1	5
3	10
	15
4	

6.

Number of People	Entrance Fee
2	10
4	20
	30
8	
	50

Fill in the tables to help you solve these problems.

7. Maria bought 3 apples for 60¢. Paul spent 80¢ for 4 apples. Connie got 5 apples from the store for $1.00.

Number of Apples	Cost
3	60¢
4	80¢
5	

a. If Zelda wanted to buy 8 apples, how much would she have to spend?

b. If Sarah has $1.70, can she buy 9 apples?

c. Xavier's aunt, uncle, and cousin each wanted 4 apples. How much money does he need?

8. To evaluate the expression $x + 3$, [STO▶] the number in the chart for [XIT]. [XIT] [+] [3] [ENTER] will evaluate $x + 3$.

Record the results in the second column. The first one is done for you.

Number	Result
8	11
9	
10	
11	
12	
13	

9. Using the same procedure, evaluate the expression "3 times x." You may pick any number to use in the first column.

Number	Result

10. Write an algebraic expression for questions 1–6. Verify with the graphing calculator.

a. Table 1

b. Table 2

c. Table 3

d. Table 4

e. Table 5

f. Table 6

Additive Inverses and Identities

Using "Special Addition Numbers" students will

- Identify the corresponding additive inverse for given values and discover the law of addition of inverses
- Find the additive identity for given values and discover the law of additive identity
- Review decimal and fraction equivalents

Teaching the Lesson

- While the properties of the concept are explored in this lesson, it is important to pay close attention to the vocabulary and terms. Note the difference between the [(–)], negative key, and the [–], subtract key. Use [(–)] to find the negative of G, –G.

Special Addition Numbers

1. Complete the table by recording the results of your operations. Use the negative key, [(–)], to find the negative of G, –G.

G = given value	–G	G + (–G)	G + 0
5			
8			
16			
32			
–5			
–8			
–16			
–32			

2. Is there a pattern in each column? Describe them.

   –G column.

   G + (–G) column.

   G + 0 column.

3. Why is 0 a special number in addition? Explain.

4. In the –G column, how does the – sign affect the original value?

5. What would happen if the given value was a variable, like x or a?

6. Which do you think is the additive identity? Which column demonstrates this? Why is it called that?

7. The pattern in G + (–G) column demonstrates the property of *additive inverse*. G and –G are inverses or opposites. In your own words, describe this property.

8. Predict the value of each expression below with the laws you have defined. Use your calculator to test your prediction. (Use the [(–)] for negative of and [–] for subtraction.)

a. 5 + (–5) = ______ b. –8 + 8 = ______ c. 16 + 0 = ______

d. 0 + 32 = ______ e. –5 + (–5) = ______ f. 0 + (–8) = ______

g. –16 + 0 = ______ h. 0 + (–32) = ______ i. a + 0 = ______

j. –b + 0 = ______

Multiplicative Identities, Inverses, and More

Using "Special Multiplication Numbers" students will

- Identify the corresponding multiplicative inverse for given values
- Discover the law of multiplication of inverses
- Find the multiplicative identity for given values and discover the law of multiplicative identities
- Recognize that multiplication by zero produces zero and discover the law of multiplication by zero
- Discover that multiplication by –1 produces the opposite of a given value and discover the law of multiplication by –1

Teaching the Lesson

- G^{-1} is the multiplicative inverse. Use [X^{-1}] to find G^{-1}.
- $G \times -1$ is multiplying by negative one. Use [(–)] to multiply by negative one.
- While the properties and concepts are thoroughly explored, careful attention to the vocabulary is important.

Special Multiplication Numbers

1. Complete the table by recording the results of your operations. Use the edit commands to rewrite and edit the prior expression. Use the $\boxed{X^{-1}}$ key to find G^{-1}.

G = given value	G^{-1}	$G \times G^{-1}$	$G \times 1$	$G \times -1$	$G \times 0$
5					
8					
16					
32					
–5					
–8					
–16					
–32					

2. Is there a pattern in each column? Describe each pattern.

   a. G^{-1} column

   b. $G \times G^{-1}$ column

   c. $G \times 1$ column

   d. $G \times -1$ column

   e. $G \times 0$ column

3. Why are 0, 1, and –1 special numbers in multiplication? Explain.

4. What would happen if the given value was a variable, like x or a and you multiplied by 0? by 1? by –1?

5. Which do you think is the *multiplicative identity*? Which column demonstrates this? Why is it called that?

6. The pattern in G × 0 column demonstrates the *multiplication property of zero*. In your own words, describe this property.

Reciprocals as Inverses

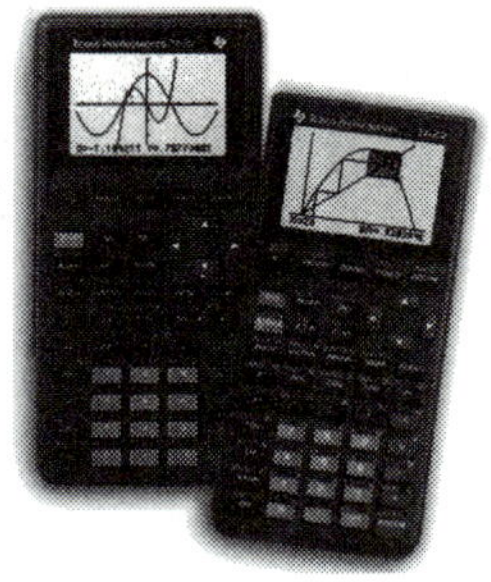

Using "Topsy-Turvy" students will

- Identify the corresponding inverse for given values
- Identify the corresponding reciprocal for given values
- Recognize the reciprocal and the multiplicative inverse are the same

Teaching the Lesson

- G^{-1} refers to the multiplicative inverse; the calculator will give a decimal value.
- Students may need help recognizing that $R = G^{-1}$ for the fractional values of G.
- Use [X^{-1}] to find G^{-1}.

Topsy-Turvy

1. Complete the table by recording the results of your operations. Find the reciprocal by inverting the fraction. Use the $\boxed{X^{-1}}$ key to find G^{-1}.

G = given value	*R* = Reciprocal	G^{-1} = Inverse	$G \times R$	$G \times G^{-1}$
5				
8				
16				
32				
–5				
–8				
–16				
–32				
$\frac{1}{5}$				
$-\frac{1}{5}$				
$\frac{7}{8}$				
$\frac{5}{32}$				
$-\frac{5}{32}$				
$\frac{1}{2}$				

2. Look at the G^{-1} and the *R* columns. Are there any similarities in the two?

3. Compare the $G \times R$ and the $G \times G^{-1}$ column. What do you notice?

4. G^{-1} has been identified as a special number. What happens when you multiply G by its multiplicative inverse, G^{-1}?

5. Look at the $G \times R$ column. What happens when you multiply G by its reciprocal, R?

6. In multiplication, is it possible to have two numbers that act the same? What must be true about G^{-1} and R?

Comparing and Ordering Fractions

Using "Least to Greatest and Greatest to Least" students will

- Use the TEST menu to predict and verify the relative size of fractions
- Compare and order sets of fractions

Teaching the Lesson

- This lesson uses features found only on the TI-82.
- Clear the L_1 data list with ClrList in the STAT menu.
- When using the TEST menu you get a binary result: 1 indicates a true result, and 0 indicates a false result.
- Knowledge of the SortA and SortD options of the LIST menu as well as the ▶ Frac option of the MATH menu are important to this lesson.

Least to Greatest and Greatest to Least

1. Predict using >, <, or =.

   a. $\frac{3}{4}$ _____ $\frac{2}{3}$ b. $\frac{5}{8}$ _____ $\frac{4}{7}$ c. $\frac{6}{11}$ _____ $\frac{12}{23}$

   d. $\frac{43}{53}$ _____ $\frac{203}{25}$ e. $\frac{93}{200}$ _____ $\frac{111}{250}$ f. $\frac{133}{255}$ _____ $\frac{798}{1530}$

2. Verify your predictions using the TEST menu. Enter the left side of the expression using the ÷ to represent the fraction bar, select your prediction from the TEST menu, followed by the right side of the expression using the ÷ to represent the fraction bar, then ENTER. Remember, a result of 1 indicates a correct prediction, and a result of 0 indicates an incorrect prediction. Make any corrections that are necessary.

3. Order the following fractions from least to greatest. Compare any two at a time using the TEST menu.

   $\frac{739}{2000}, \frac{9}{25}, \frac{923}{2500}$ ___ ___ ___

4. Now order the following set of fractions from largest to smallest.

   $\{\frac{3}{4}, \frac{5}{8}, \frac{2}{5}, \frac{4}{5}, \frac{4}{9}, \frac{3}{8}, \frac{6}{7}\}$ {___ ___ ___ ___ ___ ___ ___}

Ordering the set by comparing two fractions at a time can be tedious. To rapidly order the set, use the statistical capabilities of the TI-82. First enter your data, then sort it. To enter the data, access the STAT menu and Edit the data into L_1. To order the from greatest to least, use the STAT menu to SortD and ENTER. Place L_1 after the SortD command to indicate that you want to sort L_1 in descending order and ENTER.

5. To view the sorted data, select L_1, then from the MATH menu, select ▶ Frac to insure that the data is in fraction form and ENTER. Use the right and left arrows to scroll through the data. Record the sorted data.

   {_____, _____, _____, _____, _____, _____, _____}

6. Create a list of at least five fractions to sort in ascending order. Exchange your list with a partner. Can your partner sort your list properly? Can you sort your partner's list properly? Be sure you have sorted your own list prior to sharing it with a partner.

TOPIC 2

Working with Patterns

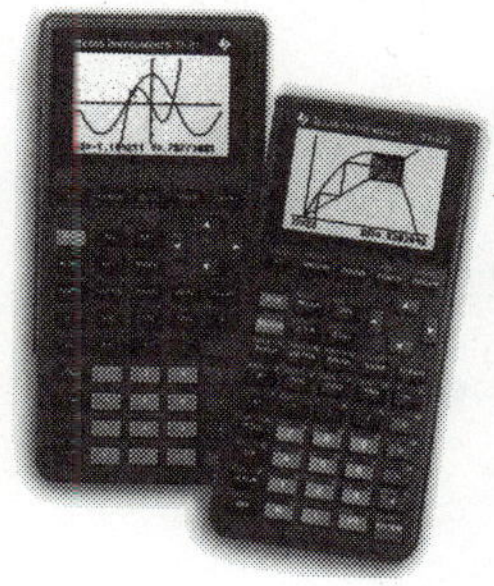

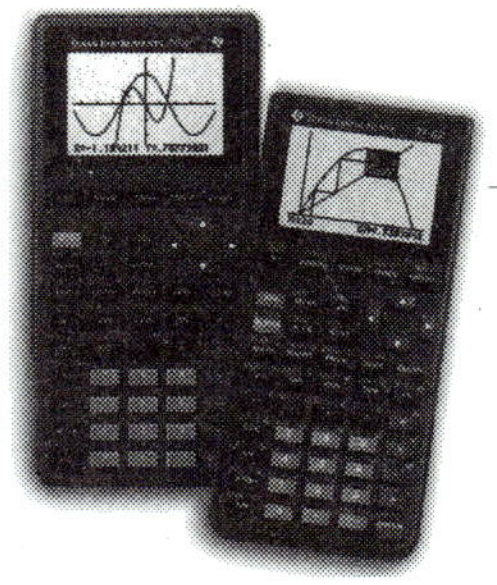

Palindromes

Using "Palindrome Patterns" students will

- Discover patterns using problem solving strategies of *guess and check* and *organizing data*
- Use the edit commands on the calculator

Teaching the Lesson

- Materials: 1–100 Chart, 101–200 Chart, colored pencils
- The patterns in the 100 Chart are very distinct if the students follow the directions asking them to color code the squares. Most numbers take no more than 2 iterations to produce a palindrome. The numbers 89 and 98 require 24 iterations before producing a palindrome. The pattern in the 101–200 Chart is not as distinctive and takes quite a long time to complete.
- The large display and edit commands on the graphing calculator are advantageous in this lesson. However, a regular calculator will suffice.
- When doing the iterations for 89, after the 18th iteration, the number of digits in the number exceeds the screen display. The remaining iterations will have to be calculated by hand.

Solution

1–100 Chart

1 (0)	2 (0)	3 (0)	4 (0)	5 (0)	6 (0)	7 (0)	8 (0)	9 (0)	10 (0)
11 (0)	12 (1)	13 (1)	14 (1)	15 (1)	16 (1)	17 (1)	18 (1)	19 (2)	20 (1)
21 (1)	22 (0)	23 (1)	24 (1)	25 (1)	26 (1)	27 (1)	28 (2)	29 (1)	30 (1)
31 (1)	32 (1)	33 (0)	34 (1)	35 (1)	36 (1)	37 (2)	38 (1)	39 (2)	40 (1)
41 (1)	42 (1)	43 (1)	44 (0)	45 (1)	46 (2)	47 (1)	48 (2)	49 (2)	50 (1)
51 (1)	52 (1)	53 (1)	54 (1)	55 (0)	56 (1)	57 (2)	58 (2)	59 (3)	60 (1)
61 (1)	62 (1)	63 (1)	64 (2)	65 (1)	66 (0)	67 (2)	68 (3)	69 (4)	70 (1)
71 (1)	72 (1)	73 (2)	74 (1)	75 (2)	76 (2)	77 (0)	78 (4)	79 (6)	80 (1)
81 (1)	82 (2)	83 (1)	84 (2)	85 (2)	86 (3)	87 (4)	88 (0)	89 (24)	90 (1)
91 (2)	92 (1)	93 (2)	94 (2)	95 (3)	96 (4)	97 (6)	98 (24)	99 (0)	100 (1)

1–100 Chart

1	2	3	4	5	6	7	8	9	10
11	12	13	14	15	16	17	18	19	20
21	22	23	24	25	26	27	28	29	30
31	32	33	34	35	36	37	38	39	40
41	42	43	44	45	46	47	48	49	50
51	52	53	54	55	56	57	58	59	60
61	62	63	64	65	66	67	68	69	70
71	72	73	74	75	76	77	78	79	80
81	82	83	84	85	86	87	88	89	90
91	92	93	94	95	96	97	98	99	100

101–200 Chart

101	102	103	104	105	106	107	108	109	110
111	112	113	114	115	116	117	118	119	120
121	122	123	124	125	126	127	128	129	130
131	132	133	134	135	136	137	138	139	140
141	142	143	144	145	146	147	148	149	150
151	152	153	154	155	156	157	158	159	160
161	162	163	164	165	166	167	168	169	170
171	172	173	174	175	176	177	178	179	180
181	182	183	184	185	186	187	188	189	190
191	192	193	194	195	196	197	198	199	200

Palindrome Patterns

A *palindrome* is a number that reads the same forward and backward. Any single-digit number is a palindrome.

Examples   21 is not a palindrome

1991 is a palindrome

You can obtain a palindrome by reversing the digits and adding the results to the original number. Repeat these steps until a palindrome result.

Examples

$$\begin{array}{r} 21 \\ +12 \\ \hline 33 \end{array}$$

One step required.

$$\begin{array}{r} 28 \\ +82 \\ \hline 110 \\ \text{reverse and add again} \quad +011 \\ \hline 121 \end{array}$$

Two steps required.

Each step required to create a palindrome is called an *iteration*.

Use your 100 Chart and your graphing calculator to create palindromes. Color the numbers

- That require 0 iterations one color,
- That require 1 iteration another color,
- That require 2 iterations a third color, and so on.

Look for patterns. What kind of patterns are formed?

1. In paragraph form, describe the patterns that you see in the chart.

2. Of the first 100 numbers, how many require 4 iterations? 6 iterations?

3. How many iterations are required before 89 becomes a palindrome?

4. Is it possible to obtain a palindrome from all the numbers less than 100?

5. Repeat the above for a 101–200 Chart. Is it possible to obtain a palindrome from all the numbers less than 200?

Patterns and Sequences

Using "Sequences" students will

- Continue sequences (visual and numerical)
- Design their own sequences (visual and numerical)
- Using the Fibonacci Program, generate a pattern
- Investigate the Fibonacci Program to analyze which portions of the program generate the different elements in the sequence
- Analyze the Fibonacci sequence and describe the pattern

Teaching the Lesson

- Copy the overhead transparency and use it to introduce sequences.
- This program is quick and easy to enter. Remember to press ENTER to generate each value in the sequence.
- Here are the command locations for the Fibonacci program.

Command	Where to Find It
the little arrow	STO key
Lbl	PRGM menu
Disp	PRGM menu, I/O submenu
Pause	PRGM menu
Goto	PRGM menu
End	PRGM menu

Solutions

Overhead transparency

3. 28, 39, 52, 67
4. perfect squares
6. 46, 61, 88, 97, 118
7. 9, 4, 1, 0, 1, 4
   (the units digits of the perfect squares)
10. 91, 140, 204, $\frac{n(n+1)(2n+1)}{6}$
   (the total number of squares in each pattern in problem 9)

Group Worksheets

2. To complete the sequence, imagine a clock face.
3. Powers of two plus one.

Continue these patterns.

1. 1, 2, 3, 4, ______, ______, ______, . . .

2. 2, 4, 6, 8, ______, ______, ______, . . .

3. 3, 4, 7, 12, 19, ______, ______, ______, . . .

4. 1, 4, 9, 16, ______, ______, ______, . . .

5. 1, 6, 13, 22, 33, ______, ______, ______, . . .

6. 2, 4, 9, 6, 5, 6, ______, ______, ______, . . .

7. 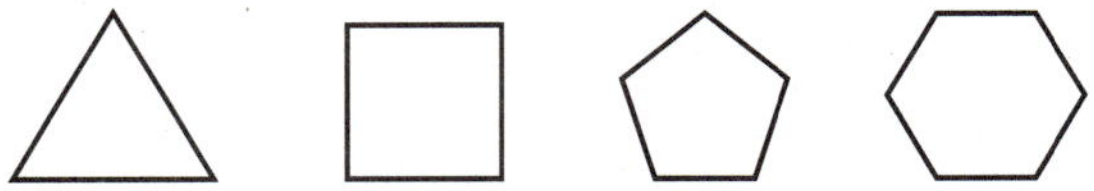 , ______, ______, ______, . . .

8. 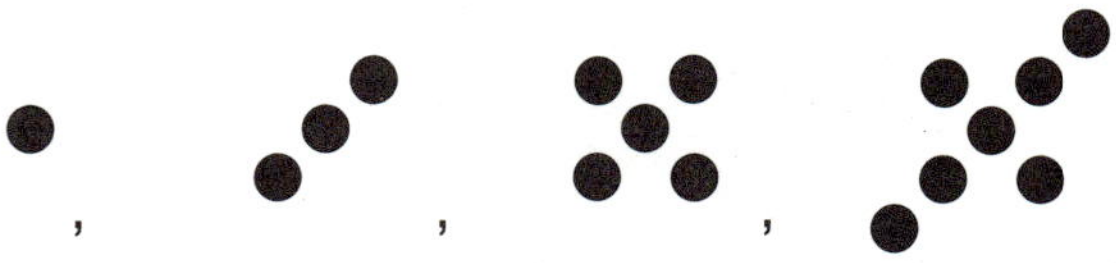 , ______, ______, ______, . . .

9. 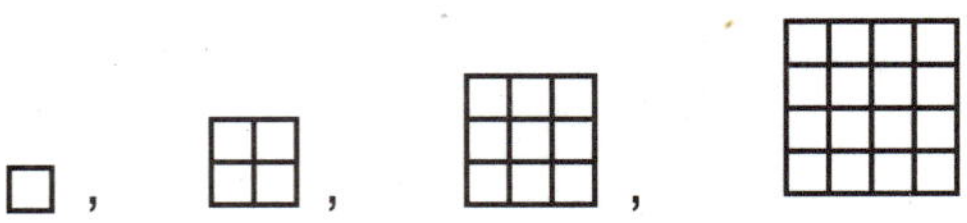 , ______, ______, ______, . . .

10. 1, 5, 14, 30, 55, ______, ______, ______, . . .

Sequences

Continue the sequences. Fill in the blanks.

1. 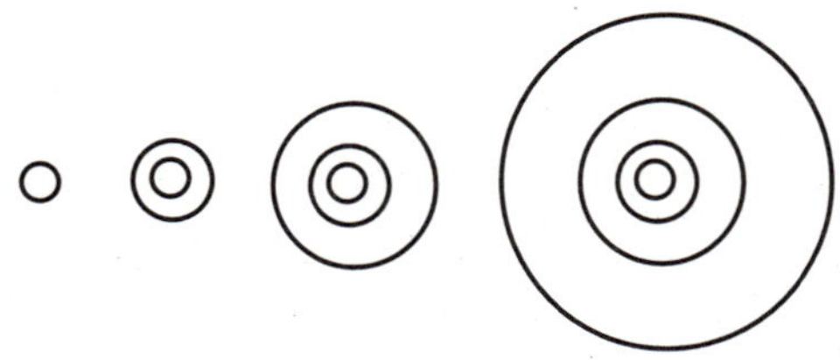 , ______, ______, ______, . . .

2. 1, 6, 11, 4, 9, 2, 7, _____, _____, _____, . . .

3. 5, 9, 17, 33, 65, _____, _____, _____, . . .

4.

A	B	C	D
			E
K			F
J	I	H	G

5. Have your group generate several sequences (using letters, numbers, pictures, etc.). Choose your favorites (two or more). Put them on an overhead transparency for use with the rest of the class.

6. Enter the following Program.

```
:0→X
:1→Y
:Lbl 1
:X+Y→A
:X→Y
:A→X
:Disp X
:Pause
:Goto 1
:End
```

7. Run the program, record the first fifteen numbers in the pattern. Remember to press ENTER after each.

8. Describe how the 13 was generated.

9. Why was 1 the only number repeated in the program?

10. What happens in the line X+Y→A?

11. Which line stores the next element in the sequence?

12. This special pattern is called the Fibonacci Sequence. Describe this pattern.

Linear Expressions

Using "Patterns with Numbers" students will

- Generate values for a table
- Generalize the *n*th term of a linear expression
- Verify their generalizations of the *n*th term using their calculators

Teaching the Lesson

- This lesson uses features found only on the TI-82.
- Before beginning this lesson, in the TblSet menu, set the following parameters:

  TblMin = 0

  ΔTbl = 1

  Indpnt: Auto

  Depend: Ask
- Use the program "Patterns from Equations" on page 149 to allow students to define a function, enter it in the program, and challenge other students to identify the function that determines the output for each number they enter.

Patterns with Numbers

1. In the [Y=] menu enter $Y_1 = X$. Use [TABLE] to investigate the relationship between x and y_1 Use your right arrow and down arrow keys to highlight and generate the values for y_1 that correspond to x. Each time you are ready for the next y_1 value, cursor down and press [ENTER]. Record the data requested.

x	y_1
0	
1	
2	
3	
10	
15	
21	

a. Is there a pattern? What is the relationship between x and y_1?

b. If x were 100, what would be the value of y_1?

c. Describe how y_1 is generated for any value of x?

2. In the [Y=] menu enter $Y_1 = 3X$. Use [TABLE] to investigate the relationship between x and y_1. Use your right arrow and down arrow keys to highlight and generate the values for y_1 that correspond to x. Each time you are ready for the next y_1 value, cursor down and press [ENTER]. Record the data requested.

x	y_1
0	
1	
2	
3	
10	
15	
21	

a. Is there a pattern? What is the relationship between x and y_1?

b. If x were 100, what would be the value of y_1?

c. Describe how y_1 is generated for any value of x?

3. In the [Y=] menu, enter $Y_1 = 2X + 3$. Use [TABLE] to investigate the relationship between x and y_1. Use a right arrow and down arrow keys to highlight and generate the values for y_1 that correspond to x. Each time you are ready for the next y_1 value, cursor down and press [ENTER]. Record the data requested.

x	y_1
0	
1	
2	
3	
10	
15	
21	

a. Is there a pattern? What is the relationship between x and y_1?

b. If x were 100, what would be the value of y_1?

c. Describe how y_1 is generated for any value of x?

4. Generate *general terms* for each of the expressions in questions 1–3. This means

a. Using your information from the table in question 1, if x were n, what would be the value of y_1?

b. Using your information from the table in question 2, if x were n, what would be the value of y_1?

c. Using your information from the table in question 3, if x were n, what would be the value of y_1?

5. For each of the following tables, complete the missing information. Use your calculator to verify your general term by making it the value of y_1.

x	y_1
0	0
1	
3	15
5	25
10	
15	
n	

x	y_1
0	3
1	4
2	
3	6
4	
5	8
n	

x	y_1
0	1
1	4
2	
3	10
4	
5	16
n	

x	y_1
0	
2	6
4	14
6	22
8	
10	38
n	

x	y_1
0	
1	0
2	
3	2
4	
5	4
n	

x	y_1
1	
3	3
5	7
7	11
9	
11	
n	

TOPIC 3

Expressions and Equations

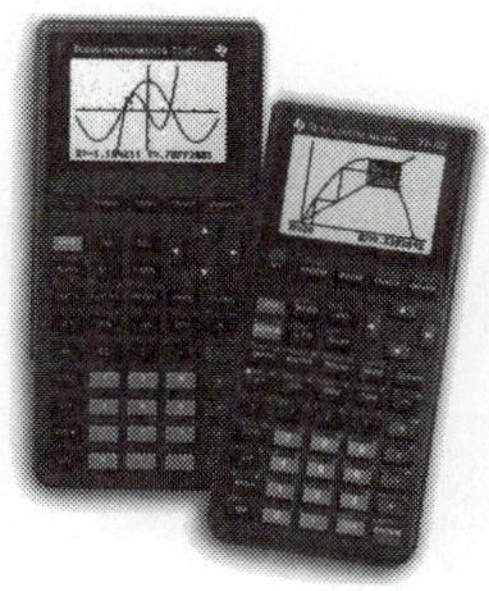

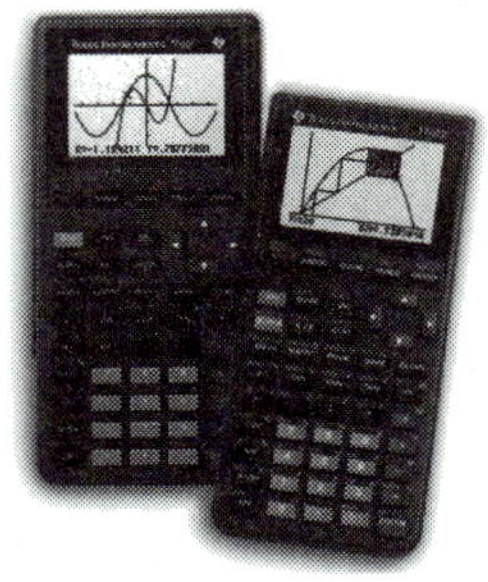

Modeling Expressions

Using "Sense About Cents" students will

- Create tables of data as a tool for problem solving
- Analyze and interpret the data to obtain a solution
- Look for patterns in their tables and learn to write expressions that model these patterns
- Use calculators to evaluate these expressions for specific values of the variables
- Learn to rewrite expressions in terms of only one variable
- Use the graphing features to graph, explore, and interpret these equations

Teaching the Lesson

After completing "Sense About Cents"

- Help students write an expression for each problem. It will be easier to break the problem into smaller expressions first.

  Example from the first problem,

  R = number of words used correctly
  W = number of words used incorrectly
  $5R$ = amount of money earned in cents
  $10W$ = amount of money lost in cents
  $5R - 10W$ = net amount of money gained or lost

- Have students clear their home screen and store their values from the table in the appropriate variables.

  Example, 18 [STO▶] R ENTER

  9 [STO▶] W ENTER

  Note the calculator confirms the value being stored.

- Have students type in the expression for the net amount of money on their home screen. Note that [ENTER] evaluates the expression.

  Explain to students that by putting in an imaginary = sign between the expression and the returned value, we get an equation.

- You may want to graph each equation so that students can see a picture and use the [TRACE] feature to find other values. To do this, students must rewrite the equation so that Y_1 equals a function of R. Students will need to rewrite their equation so that one variable is in terms of the other.

Example, rewriting $5R - 10W$ = net gained or lost from the first problem

R = number of words used correctly

$27 - R$ = number of words used incorrectly

$5R - 10(27 - R)$ = net amount gained or lost. Let this be Y_1

- Clear the [Y=] menu, and enter this equation for Y_1. Now you need to set the [RANGE] (for the TI-82 use [Window]). Note the y-axis is the net amount of money, the x-axis is the number of words spelled correctly. Under the [RANGE] menu, enter appropriate range values. The following "magic screen" values are appropriate for the first problem.

  $X_{min} = 0$
  $X_{max} = 23.75$
  $X_{scl} = 1$
  $Y_{min} = -152$
  $Y_{max} = 100$
  $Y_{scl} = 10$
  $X_{res} = 1$

- [GRAPH] and explore the line using [TRACE] and the arrow keys.

Sample Questions

1. At what point does y = 0? What does this means with respect to Peter's vocabulary words?
2. In relation to the graph, when is Peter making money? When is he losing money?
3. Why is it only necessary to TRACE values from 0 to 27?
4. What is the maximum amount of money Peter can make in one night? Also, ask students to analyze their graphs with respect to the problem.

- Using this equation, other values can also be verified by STOring the value in x, calling for Y_1 in the [Y-VARS] menu, and entering to evaluate for that value of x. Example: If Peter only got 5 words correct then $x = 5$.

  [5] [STO▶] [X] [ENTER]

  [Y-VARS] [ENTER] [ENTER]

  yields a value of –195

  so Peter has lost $1.95

Sense About Cents

Problem-Solving Hints

- Read each problem carefully.
- What does the question ask you to find?
- List the necessary facts.
- Estimate some reasonable values for your answer.
- Create a table (or tables) to help organize your information.
- Use a calculator, if necessary, to help fill in the tables.
- Study the information gathered to see if you can answer the question.
- Construct equations to solve the problems and verify your results.

1. Rebecca has 15 problems for homework. She solved 8 of them correctly, and for each correct solution she received 12¢. But for each of the 7 wrong answers, Rebecca was charged 15¢. How much did she earn?

2. Elena buys newspapers for 12¢ each and sells them for 20¢ each. Today Elena bought 50 papers, but because of rain, she only sold 22. How much money did Elena make?

3. Use the table that follows to help you answer these questions. Peter has 27 spelling words to use in sentences. His mother rewards him with a nickel for each word he uses correctly and fines him a dime for each word he misuses. Tonight, Peter neither made nor lost money. How many words did he use correctly?

a. What are you asked to find?

b. What are the given facts?

c. Predict if Peter is going to use more words correctly or incorrectly.

d. Complete this table.

Number Correct	Money Earned	Number Wrong	Money Lost
0		27	
1		26	
2		25	
3		24	
4		23	
5		22	
6		21	
7		20	
8		19	
9		18	
10		17	
11		16	
12		15	
13		14	

Number Correct	Money Earned	Number Wrong	Money Lost
14		13	
15		12	
16		11	
17		10	
18		9	
19		8	
20		7	
21		6	
22		5	
23		4	
24		3	
25		2	
26		1	
27		0	

4. How did you calculate the Money Earned column?

5. How did you calculate the Money Lost column?

6. How many words did he use correctly?

7. Tonight's science assignment consisted of 32 identifications. Since Joe does poorly in science, his father rewards him with 15¢ for each correct identification and penalizes him 25¢ for each incorrect identification. If Joe breaks even, how many identifications does he have incorrect?

8. For each box of candy Sandy sold, she earned 60¢. However, she paid $1.00 for each box she didn't sell. If she ordered 40 boxes of candy, but only broke even, how many were unsold?

Like Terms

Using "Like Terms" students will

- Learn how to [Test] for equality on the calculator
- Use the 0 and 1 test criteria to determine whether or not an expression is correct
- Answer questions regarding variables and their coefficients
- Find missing coefficients and variables
- Combine like terms

Teaching the Lesson

- The equal sign is in the [TEST] menu. Storing a 1 for *x* replaces the previously stored value of 0 for *x*. If 0 is stored, all combined term expressions will give a true result for all problems. Students may need some help storing values.

Like Terms

1. Remember, two apples plus three apples equals five apples. We want to look at this idea in another way. The apple can be represented by an x or any other symbol.
Enter the following into your calculator.

   Store 1 in X. Press [1] [STO▶] [X]. Test each of the following.

   a. $2x + 3x = 5x$

   What does the calculator display? Why?

   b. $12x - 3x = 9x$

   What does the calculator display? Why?

   c. $3x - 4x = 7x$

   What does the calculator display? Why?

   d. $3x + 4x = 7x$

   What does the calculator display? Why?

2. Explore with your calculator. Remember:

   the calculator returns a **1** for a *true* statement

   the calculator returns a **0** for a *false* statement

	What does the calculator return?	Is the equation True or False?
a. $5x + 6x = 11x$	0/1	T/F
b. $4x - 7x = 11x$	0/1	T/F
c. $9x + 8x = 18x$	0/1	T/F
d. $8x - 9x = x$	0/1	T/F
e. $7x + 8x = 15x$	0/1	T/F

3. $12x + 34x = ?$

a. Predict the answer.

b. Write a description of how you arrived at your answer.

c. Test your prediction with the calculator.

4. Without a calculator, complete the following, then verify your solutions with the calculator.

a. $12x + 6x =$ ________

b. $14x - 5x =$ ________

5. Each product of a coefficient and a variable is called a term. Terms with the same variables are called like terms. Fill in the blank with the term that makes the statement true. Use your calculator to test your solution.

a. $3x +$ _____ $+ 2x = 9x$

b. $15x - 3x -$ _____ $= 7x$

c. $12x + 4x - 2x +$ _____ $= 21x$

d. $23x + 16x -$ _____ $= 32x$

e. $36x - 4x +$ _____ $= 48x$

f. $2x +$ _____ $+ x = 6x + x$

g. $15x - 6x +$ _____ $= 14x$

Solving Equations I

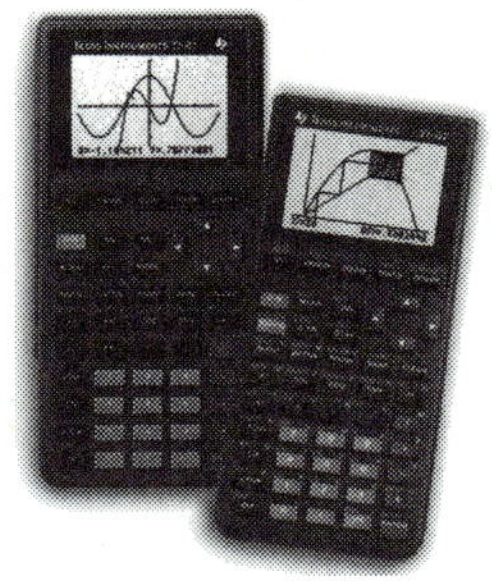

Using "Guess, Check, Revise" students will

- Evaluate an equation for a given value
- Enter the numeric expression in the calculator
- Use the edit commands to revise a guess and record the results in a chart
- Substitute values into an equation and analyze the results to make a better prediction

Teaching the Lesson

- Using the edit commands will enable students to make many predictions in a short time, allowing them to become more accurate in their predictions. There are natural breaking points between pages. Students become more accurate in predicting when given the variety of examples included in the entire lesson.
- Ask students to verbalize how they decided on their first guess and how they revised each guess.

Guess, Check, Revise

Find a solutions by guessing, checking, analyzing, and guessing again when necessary. Complete the chart for each problem.

1. $2x - 14 = 55$ Solution _______

Guess	$x = ?$	Check	Analyze
1st			
2d			
3d			
4th			
5th			
6th			

2. $68 + 42t = 1076$ Solution _______

Guess	$t = ?$	Check	Analyze
1st			
2d			
3d			
4th			
5th			
6th			

3. $12y - 9 = -45$ Solution _______

Guess	$y = ?$	Check	Analyze
1st			
2d			
3d			
4th			
5th			
6th			

4. $92z + 61 = -675$ Solution _______

Guess	$z = ?$	Check	Analyze
1st			
2d			
3d			
4th			
5th			
6th			

5. Analyze questions 1–4.

a. How did the results of questions 1 and 2 differ from the results of questions 3 and 4?

b. How can you tell from the equation whether the solution should be positive or negative?

6. $16d + 82 = 90$ Solution _______

Guess	d = ?	Check	Analyze
1st			
2d			
3d			
4th			
5th			
6th			

7. $15v + 9 = 21$ Solution _______

Guess	v = ?	Check	Analyze
1st			
2d			
3d			
4th			
5th			
6th			

8. $39 + 55w = 17$ Solution _______

Guess	w = ?	Check	Analyze
1st			
2d			
3d			
4th			
5th			
6th			

9. $180g - 31 = 45$ Solution _______

Guess	g = ?	Check	Analyze
1st			
2d			
3d			
4th			
5th			
6th			

10. Analyze questions 6–9.

a. How were your solutions different from those in questions 1–4?

b. How can you tell from the equation, when the solution might be a decimal? Explain.

11. $\frac{x}{2} + 51 = 150$ Solution _______

Guess	$x = ?$	Check	Analyze
1st			
2d			
3d			
4th			
5th			
6th			

12. $\frac{112}{y} - 22 = -36$ Solution _______

Guess	$y = ?$	Check	Analyze
1st			
2d			
3d			
4th			
5th			
6th			

13. $31 + \frac{48}{w} = 51$ Solution _______

Guess	$w = ?$	Check	Analyze
1st			
2nd			
3rd			
4th			
5th			
6th			

14. $-\frac{z}{2.1} + 2 = 1$ Solution _______

Guess	$z = ?$	Check	Analyze
1st			
2nd			
3rd			
4th			
5th			
6th			

15. Analyze questions 11–14.

a. Is there a way to predict whether your first guess should be positive or negative? How?

b. Is there a way to predict whether your guess should be a decimal? How?

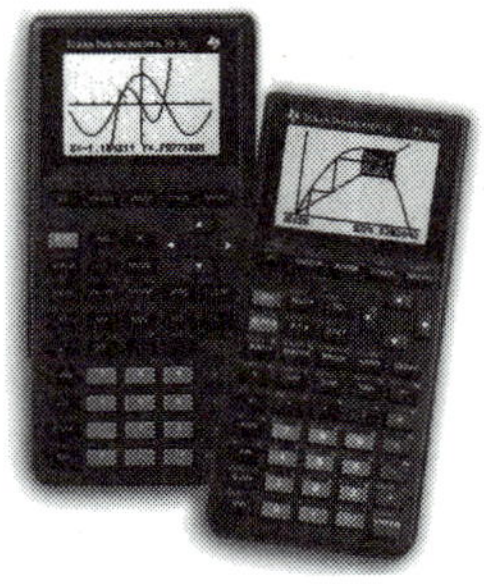

Solving Equations II

Using "Find the Intersection" students will

- Use the graphing capabilities of the calculator
- Graph expressions
- Solve the equation by finding the point of intersection
- Verify the solution by substitution on the calculator

Teaching the Lesson

- $\boxed{Y_1=}$ will always hold the left side of the equation.
- $\boxed{Y_2=}$ will always hold the right side of the equation.
- Clear all previous equations from the $\boxed{Y=}$ menu before graphing. Tracing may change the range values in your screen setting. Have students check the range setting after each problem and reset to the Suggested Range before entering the equations for the next problem. Using the Suggested Range will give integer values when tracing.

Find the Intersection

	TI-81	TI-82
Suggested Range	$X_{min} = -47$	$X_{min} = -46$
	$X_{max} = 48$	$X_{max} = 48$
	$X_{scl} = 1$	$X_{scl} = 1$
	$Y_{min} = -32$	$Y_{min} = -31$
	$Y_{max} = 31$	$Y_{max} = 31$
	$Y_{scl} = 2$	$Y_{scl} = 2$
	$X_{res} = 1$	

1. To solve $x + 5 = 11$ Enter Y_1 as the left side of the equation; that is, $x + 5$. Enter Y_2 as the right side of the equation; that is, 11. Graph.

   TRACE to find the point of intersection for the two lines. Write the coordinate pair for the point of intersection. $(x, y) =$ (_______, _______)

   What do you notice about the values of x and y at the point of intersection?

   Check the solution for x.

2. What is meant by a solution of an equation?

3. How can the solution to an equation be obtained using graphing?

4. How do you check (verify) the solution of an equation?

Solve these equations in the same manner. Be sure to verify your solutions.

5. $x - 3 = -10$
6. $x - 2 = -7$
7. $2x = 14$
8. $5x = -40$
9. $\frac{x}{2} = 9$
10. $\frac{x}{3} = -4$
11. $4x + 7 = 15$
12. $6x + 2 = -4$
13. $3x - 4 = 8$
14. $5x - 6 = -36$
15. $2x + 5 = 7x - 10$
16. $2(x + 3) = -3(x + 2)$

Equations and Problem Solving

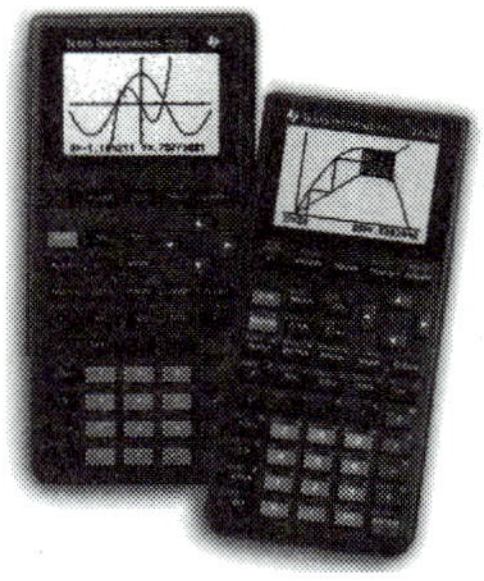

Using "Popcorn" students will

- Set up equations for cost, revenue, and profit
- Use the graphing capabilities to graph the above equations
- Use the TRACE feature to investigate the equations
- Investigate the change in profit as selling price changes
- Conduct an investigation to determine the best selling price

Teaching the Lesson

- Clear all previous equations from the Y= menu before starting. Students may need help setting up the cost, revenue, and profit equations. When using the calculator, remind the students that to de-select Y_1 or Y_2, move the cursor to the = sign and press ENTER. To re-select, do the same. Tracing will give values for x and y at that particular point on the graph.
- After student, individually or in groups, have completed questions 1-15, ask them to design a Popcorn Sales Report. Copy these instructions on a transparency or make a copy for each group

  Design a Popcorn Sales Report stating the amount of profit that could be earned for various selling prices. Start with 10¢ and using an increment of 10¢, determine the profit up to a sales price of a $1.00. A complete report would include graphs and an explanation of how you arrived at your conclusions. To compete your sales report, survey your class or school to discover what price they would be willing to pay. Then make a recommendation as to the best selling price, including why you believe this is the best price.

Popcorn

Your class decides to sell bags of popcorn as a fund raiser. The cost to rent the popcorn machine is $15.00, and it cost 10¢ a bag to produce the popcorn. The class wants to sell the popcorn for 25¢ a bag. Using this information, complete the chart.

Number of Bags of Popcorn	Cost (Y_1) =	Revenue (Y_2) =	Profit (Y_3) =
0			
1			
10			
20			
130			

1. Write the equation for determining the cost. Cost = ____________________

2. Write the equation for determining the revenue. Revenue = ____________________

3. Write the equation for determining the profit. Profit = ____________________

4. Suggested Range
   $X_{min} = 0$
   $X_{max} = 950$
   $X_{scl} = 10$
   $Y_{min} = -20$
   $Y_{max} = 43$
   $Y_{scl} = 10$
   $X_{res} = 1$

   Press GRAPH to view the coordinate grid.
   What does the *x*-axis represent?
   What does the *y*-axis represent?

5. Enter your cost equation as Y_1, and press GRAPH.
   What does the *x*-axis represent?
   What does the *y*-axis represent?

6. Enter your revenue equation as Y_2, and press **GRAPH**. (De-select Y_1)
What does the x-axis represent?
What does the y-axis represent?

7. Enter your profit equation as Y_3, and press **GRAPH**. (De-select Y_1 and Y_2)
What does the x-axis represent?
What does the y-axis represent?

8. Graph all three equations simultaneously. What do the points of intersection tell you?

9. Why are there three lines on your graph?

10. What does each line represent?

12. What is the least number of bags of popcorn that must be sold to make a profit? How did you find this value?

13. Fill in the following table using the equations in Y_1, Y_2, and Y_3. Use TRACE to arrive at the proper x value.

Bags of Popcorn (x0	Cost (Y_1) =	Revenue (Y_2) =	Profit (Y_3) =
15			
29			
73			
129			
180			
		$20.4C	
		$24.0C	
	$13.50		
	$31.50		
			$–9.60
			$3.90
			$6.60

14. If the class wanted to make $100.00 profit, how many bags of popcorn do they have to sell? How did you arrive at this value?

15. If students decided to change the selling price to 50¢ then how many bags of popcorn would have to be sold to make a profit? How did you arrive at this value?

Open and Closed Sentences

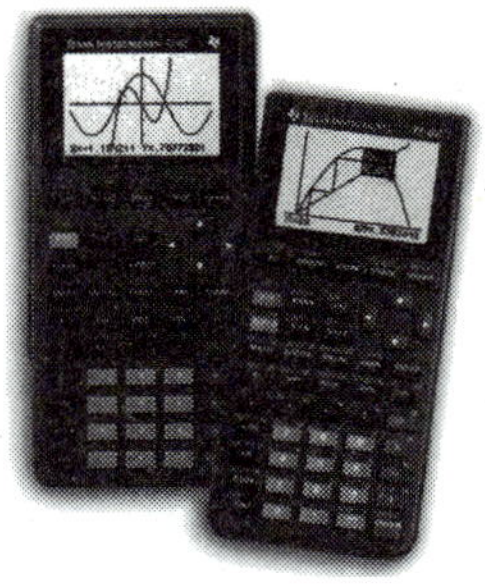

Using "Open Sentences" students will

- Find values that satisfy inequalities
- Identify endpoints
- Determine if endpoints are inclusive or exclusive
- Use the graphing capabilities
- Investigate inequality graphs using the [TRACE] feature to identify solution sets

Teaching the Lesson

- The following features of the TI-81 or TI-82 are used extensively.
- [RANGE], [Y=], [TRACE], [TEST], and [CLEAR]
- [TEST] will allow you to choose =, ≠, ≤, ≥, <, >, etc.
- When using [TRACE], Y = 0 means "false for this value of x."
  Y = 1 means "true for this value of x."
- Clear all equations from the [Y=] menu before beginning the lesson.

Open Sentences

1. Enter $Y_1 = x < 15$ using the: TEST menu. Set the MODE to DOT.

   Suggested Range
   $X_{min} = -4$
   $X_{max} = 18$
   $X_{scl} = 1$
   $Y_{min} = -2$
   $Y_{max} = 5$
   $Y_{scl} = 1$
   $X_{res} = 1$

2. GRAPH the inequality and TRACE. Locate the value for $x = 15$.

3. Is 15 a true value for x?

4. Check negative values of x. What happens to y?

5. Will it ever be 0?

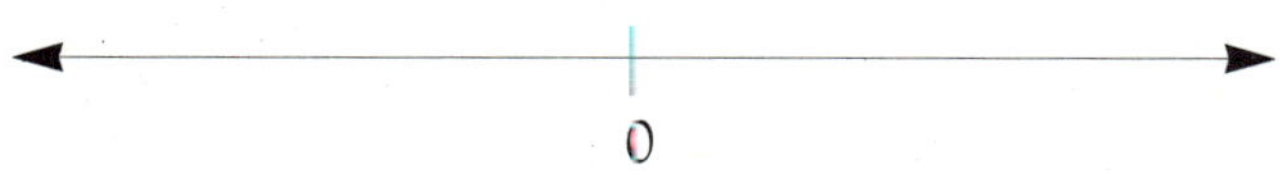

6. Draw the graph of $x < 15$.

7. In your own words, describe which values satisfy this inequality. Why?

8. Describe what occurs at the endpoint of this inequality. Why?

9. Is this an *open* or closed *interval*? Explain.

10. Graph $x \geq -2$.

Is there an open endpoint? ______ If so, name it? ______

Is there a closed endpoint? ______ If so, name it. ______

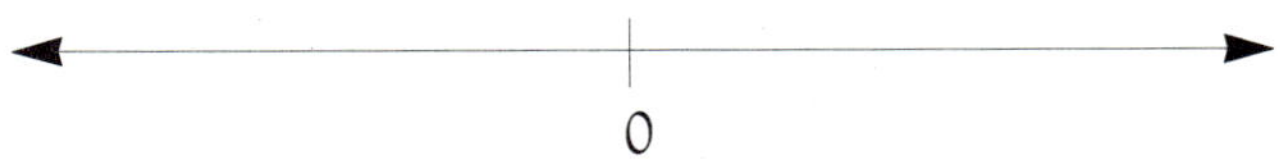

Draw the graph of $x \geq -2$.

11. Graph $x > 10$.

Is there an endpoint? ______ If so, name it. ______

Are there two endpoints? ______ If so, name them. ______

Draw the graph of $x > 10$.

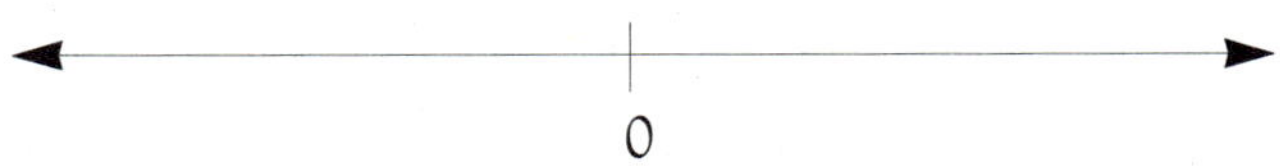

12. Graph $x \leq 4$.

Describe the endpoints of this graph.

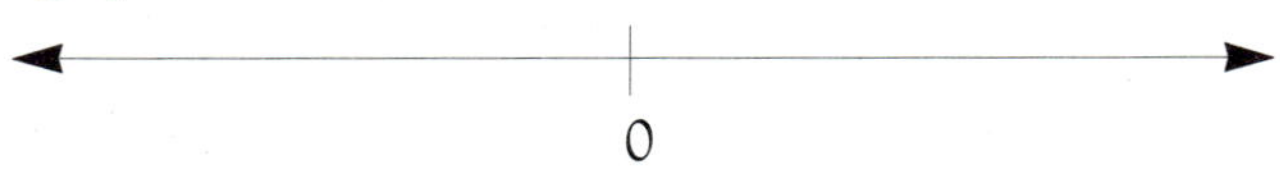

Draw the graph of $x \leq 4$.

13. To graph the compound sentence $0 < x \leq 5$, enter in Y_1, $(0 < x)(x \leq 5)$.

Describe the endpoints of this graph.

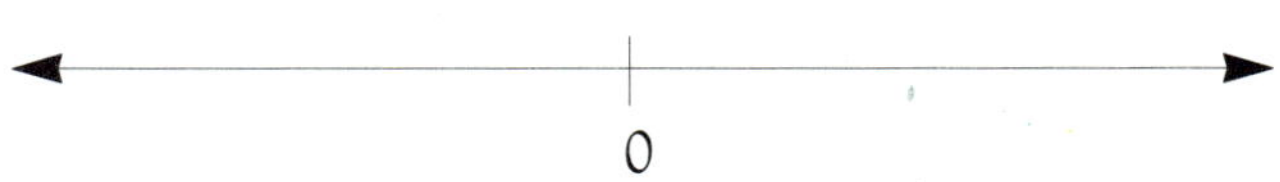

Draw the graph of $0 < x \leq 5$.

Name the endpoints of this graph.

Describe this mathematical sentence in a short paragraph.

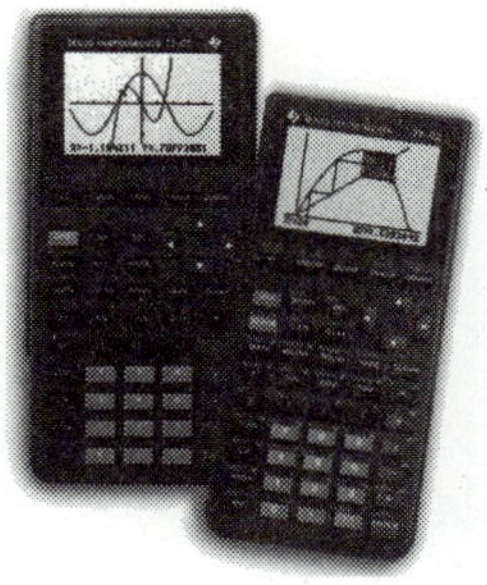

Inequality Expressions

Using "Evaluating Inequalities" students will

- Use the TEST menu to input the inequalities
- Evaluate the inequalities for the given values of x
- Describe the solutions of the inequalities
- Investigate the graphs and identify the solution sets of given inequalities using the TRACE feature
- Draw the solution sets on a number line and write as an inequality

Teaching the Lesson

- The following features of the TI-81 or TI-82 are used extensively.

  RANGE, Y=, TRACE, TEST, and CLEAR

  TEST will allow you to choose =, ≠, ≤, ≥, <, >, etc.
- Clear all previous equations from the Y= menu before beginning.
- Tracing may change the range in the viewing window. Have students check the range and re-set to the Suggested Range before entering the next equation.

1. By substituting in the given value for x, determine whether or not the following are solutions for $2x - 4 \leq 20$.

Value of *x*	Solution (Yes/No)
–20	
–2	
10	
14	
20	

2. Between what two values of x does the pattern of responses change from a yes to a no or a no to a yes?

3. Enter the following inequality for Y_1, $2x - 4 \leq 20$. Graph.

Suggested Range
$X_{min} = -4$
$X_{max} = 18$
$X_{scl} = 1$
$Y_{min} = -2$
$Y_{max} = 5$
$Y_{scl} = 1$
$X_{res} = 1$

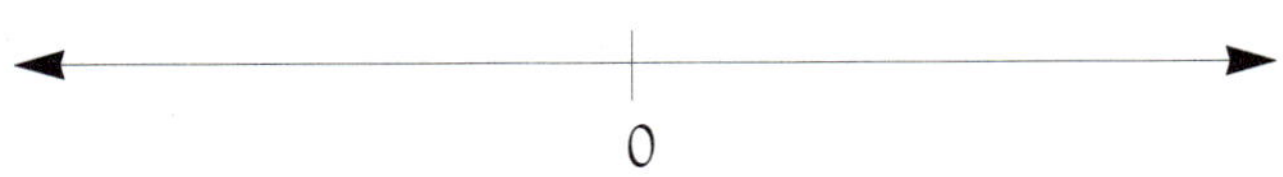

4. Graph the solution of $2x - 4 \leq 20$.

5. What conclusions can be drawn?

6. Write a solution using inequalities.

7. By substituting in the given values for x, determine whether or not the following are solutions for, $-3x + 4 > 13$.

Value of x	**Solution (Yes/No)**
–20	
–6	
–5	
–4	
–3	
–2	
–1	

8. What other values of x are solutions? What led you to this conclusion?

9. What values of x are not solutions? Why?

10. From the solutions you have listed, what can you conclude?

11. Enter the inequality for $Y_2 = -3x + 4 > 13$.

12. Identify the solution set in terms of an inequality.

13. Graph the solution of $-3x + 4 > 13$.

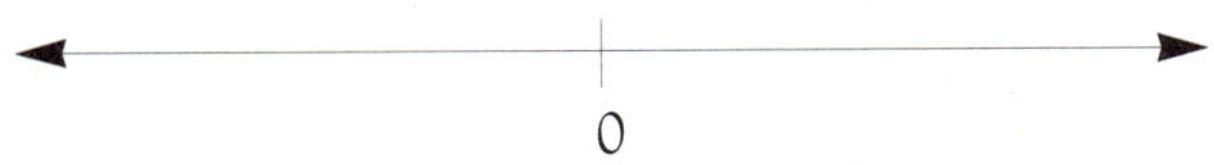

14. Identify the solution set in terms of an inequality.

15. Try the following inequalities. Identify an inequality that would describe the solution set for each one. Remember to clear each previous inequality before starting the next. Adjust the RANGE settings as necessary.

a. $24 - 5x > 8$

b. $5x \leq 42$

c. $28 \geq 2x + 6$

d. $\frac{1}{2}x + 3 < 5$

e. $-8 + x > -2$

f. $2x + 3 \geq -23$

TOPIC 4

Coordinate System and Graphing

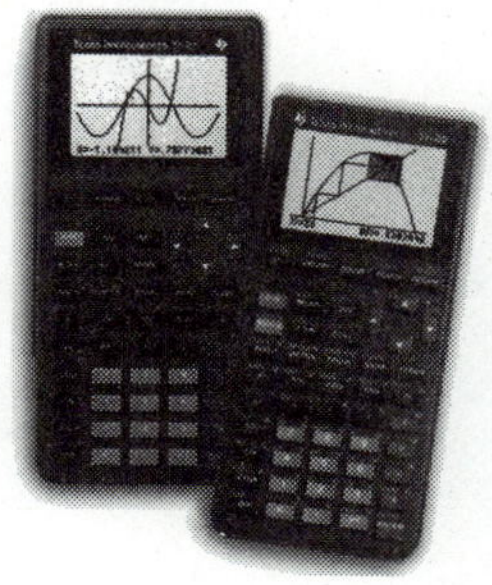

The Coordinate System and the Range Key

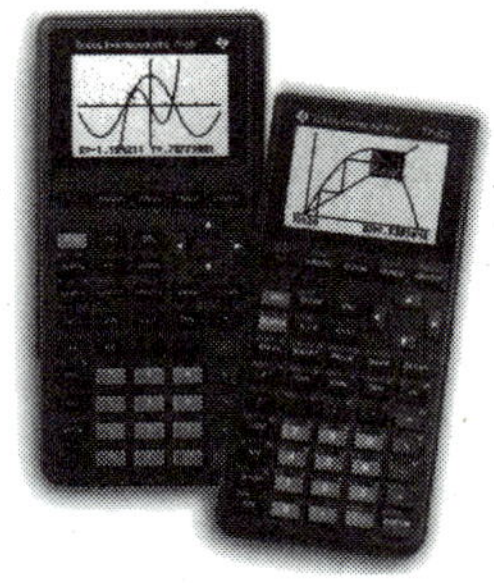

Using "Range" students will

- Use the RANGE (for the TI-82 use Window) key to determine quadrant values for *x* and *y*
- Change and explore the display of the graphing calculator
- Explore the Cartesian coordinate plane

Teaching the Lesson

- When setting the range for the viewing screen, use the negative (–), not the subtraction key.
- Clear all equations in both the Y= menu and the Draw menu before beginning.

Range

1. Press the [ZOOM] key, select the Standard setting.

a. How many marks appear on the x-axis?

b. How many marks appear on the y-axis?

2. Press [RANGE] (for the TI-82 use [Window]). Compare the values to your answers in question 1.

3. Change the X_{min} from a –10 to a –5, and the X_{max} from a 10 to a 5. Do this for the Ymin and Y_{max} values also.

4. Press [GRAPH].

a. How many marks appear on the *x* and *y* axes?

b. Is this consistent with what you saw earlier? Explain.

5. Change the values of X and Y in the RANGE to display a graph of Quadrant I. After you make your changes, press GRAPH. Did you get a graph of Quadrant I? If not, change the values until you do.

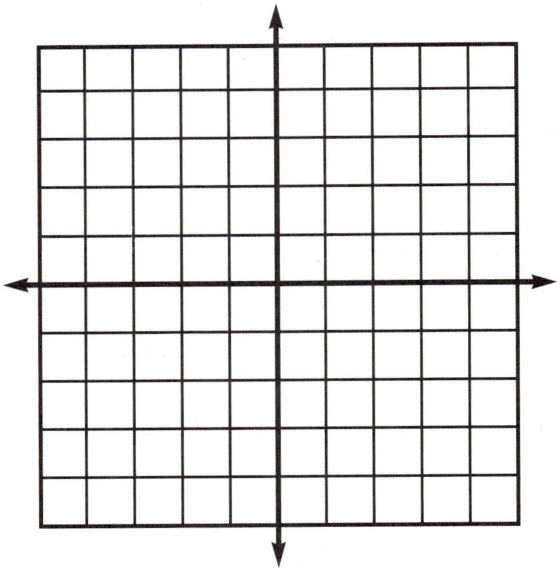

   a. What was your value for X_{min}?

   b. What was your value for X_{max}?

   c. What was your value for Y_{min}?

   d. What was your value for Y_{max}?

   e. Shade the portion of the graph displayed on the calculator on this grid.

6. Repeat Step 5 to display a graph of Quadrant II.

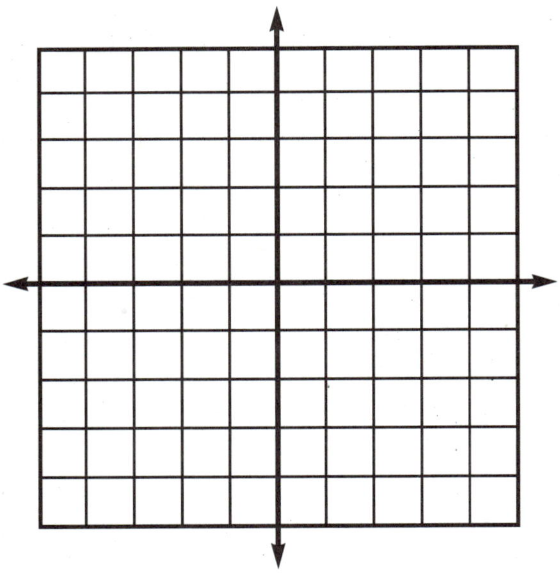

   a. What was your value for X_{min}?

   b. What was your value for X_{max}?

   c. What was your value for Y_{min}?

   d. What was your value for Y_{max}?

   e. Shade the portion of the graph displayed on the calculator on this grid.

7. Repeat Step 5 to display a graph of Quadrant III.

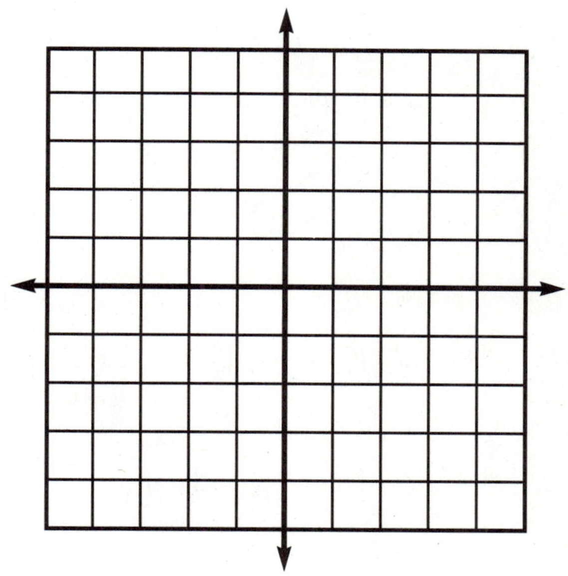

   a. What was your value for X_{min}?

   b. What was your value for X_{max}?

   c. What was your value for Y_{min}?

   d. What was your value for Y_{max}?

   e. Shade the portion of the graph displayed on the calculator on this grid.

8. Repeat Step 5 to display a graph of Quadrant IV.

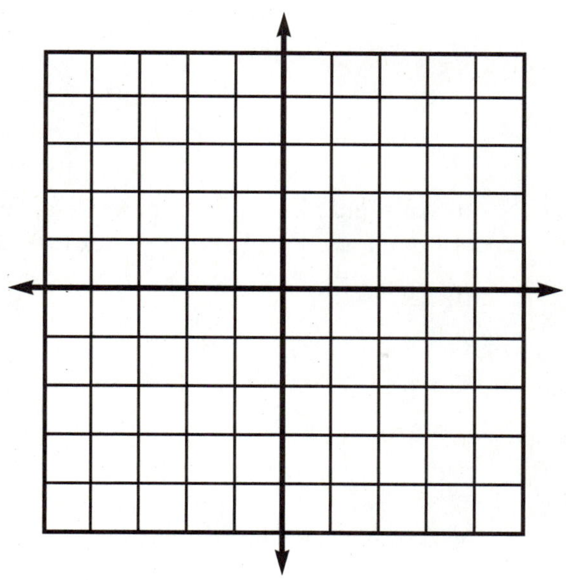

   a. What was your value for X_{min}?

   b. What was your value for X_{max}?

   c. What was your value for Y_{min}?

   d. What was your value for Y_{max}?

   e. Shade the portion of the graph displayed on the calculator on this grid.

9. What do you notice about the values used on each axis to graph each quadrant in the coordinate plane?

10. Explain how to graph each of the following using the RANGE settings.

a. Quadrants I and II only

b. Quadrants III and IV only

c. Quadrants I and IV only

d. Quadrants II and III only

e. Quadrants I and III only

11. Set the calculator back to the standard range. What value is used for X_{scl} (the x-scale)? What value is used for Y_{scl} (the y-scale)? What do these do?

12. Compare the graph using different values for the x- and y-scales. What can you conclude?

Plotting 1st Quadrant Points

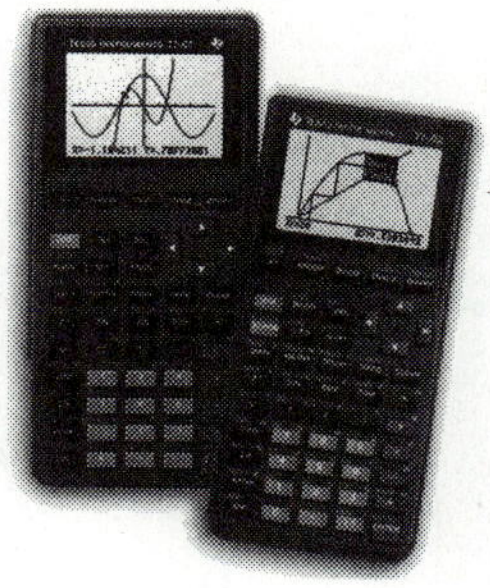

Using "Plot-a-Point" students will

- Input a series of coordinates observing the position of the plotted points
- Compare the results of plotting sets of points where one coordinate is held constant
- Look at a set of plotted points and decide which variable is held constant
- Determine the type of line formed when the *x* coordinate is constant
- Determine the type of line formed when the *y* coordinate is constant

Teaching the Lesson

- To familiarize them with the vocabulary for this lesson, discuss the following: variable, direction, constant, horizontal, vertical, point, *x*-coordinate, *y*-coordinate, abscissa, ordinate.
- The comma is an [ALPHA] character on the TI-81 and is a key character on the TI-82. Pressing [CLEAR] will take you from the grid to the Pt-On (X,Y) for easy editing.
- Make sure that both [DRAW] menu and the [Y=] menu are cleared before starting the lesson.

Plot-a-Point

To plot each ordered pair, use the Pt-On command, found under the DRAW menu on the TI-81. The Pt-On command is in the DRAW menu and Point submenu on the TI-82.

- Enter your x-coordinate.
- Enter a comma. (Remember, the comma is a ALPHA character on the TI-81.)
- Enter your y-coordinate, then close parentheses.
- Press ENTER.

1. Plot these ordered pairs (x, y) using the above steps. (Use the edit commands to save time.)

$(1, 3)$ $(2, 3)$ $(3, 3)$

$(1, 5)$ $(2, 5)$ $(3, 5)$

$(1, 7)$ $(2, 7)$ $(3, 7)$

2. Describe what you observed when x was constant.

3. Plot these ordered pairs (x, y).

$(3, 1)$ $(3, 2)$ $(3, 3)$

$(5, 1)$ $(5, 2)$ $(5, 3)$

$(7, 1)$ $(7, 2)$ $(7, 3)$

4. Describe what you observed when y was constant.

5. As you compare the points, how can you tell which coordinate value is constant in (x, y)?

6. Plot a set of points where the x-coordinate is constant. List the points below in ordered pairs (x, y).

7. Plot a set of points where the y-coordinate is constant. List the points below in ordered pairs (x, y).

8. If you connect ordered pairs (x, y) and they form a vertical line, one of the coordinates is constant. Which one is constant and why?

9. If you connect ordered pairs (x, y) and they form a horizontal line, one of the coordinates is constant. Which one is constant and why?

10. Describe the type of line formed by each of the following sets of points.

    a. (0, 1) (0, 2) (0, 5)

    b. (1, 0) (3, 0) (6, 0)

    c. (1, 3) (1, 5) (1,7)

11. Plot the following coordinate points. Describe the type of line formed.

    (1, 1) (2, 2) (3, 3) (4, 4)

Plotting Points of Linear Functions

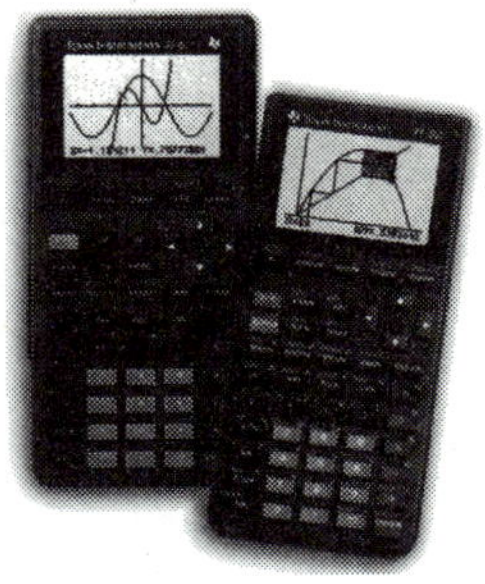

Using "Plot-a-Line" students will

- Complete a table of ordered pairs for a given function
- Set Range and enter ordered pairs as statistical data
- Plot the points using the scatter plot feature
- Plot the points on a coordinate grid
- Use the results to graph given equations
- Describe similarities and differences between functions and their graphs

Teaching the Lesson

- Clear both the [STAT] menu and the [Y=] menu before entering the data.
- The keystrokes for the TI-81 and the TI-82 are significantly different. Both are in the directions for the lesson.

Plot-a-Line

1. Consider the equation of the line $y = 2x - 1$. Use the edit commands to complete the following table.

x	$2x - 1$	y	(x, y)
5	$2(5) - 1$	9	(5, 9)
2			
1.5			
0			
–1			
–1.5			
–2			

Set an appropriate RANGE (for the TI-82 use WINDOW) using the information from the chart above.

If using the TI-81, access the STAT menu. Cursor to the DATA menu, and edit the data.

If using the TI-82, access the STAT menu, cursor to edit and enter all of the x values in L_1. Enter all of the y values in L_2, making sure that all of the values in L_1 correspond to the matching values in L_2.

Enter your data as follows.

TI-81	TI-82	
	L_1	L_2
$X_1 = 5$	$L_1(1) = 5$	$L_2(1) = 9$
$Y_1 = 9$	$L_1(2) = 2$	$L_2(2) = 3$
$X_2 = 2$	$L_1(3) =$	$L_2(3) =$
$Y_2 = 3$	$L_1(4) =$	$L_2(4) =$

Enter the data from the chart above.

To view the points on the calculator,

Using the TI-81, access the STAT menu, cursor to the DRAW submenu, and choose 2 to draw the scatter plot.

Using the TI-82, access the STAT menu, Plot 1 on scatter plot graph.

Record your scatter plot on the grid. Draw in the graph of the line $y = 2x - 1$.

What do you notice about this line?

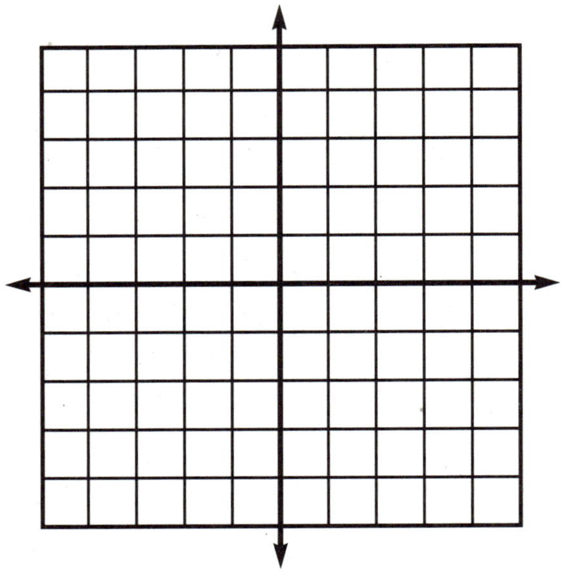

2. Repeat the steps above using the equation $y = 2x + 3$.

   a. Complete a table of ordered pairs.

   b. Graph a scatter plot for the function.

   c. Graph the equation on a coordinate grid.

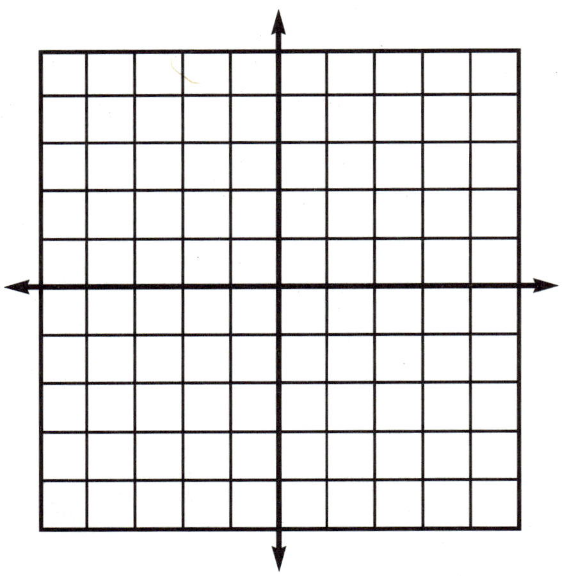

3. Repeat the steps above using the equation $y = -2x + 1$.

   a. Complete a table of ordered pairs.

   b. Graph a scatter plot for the function.

   c. Graph the equation on a coordinate grid.

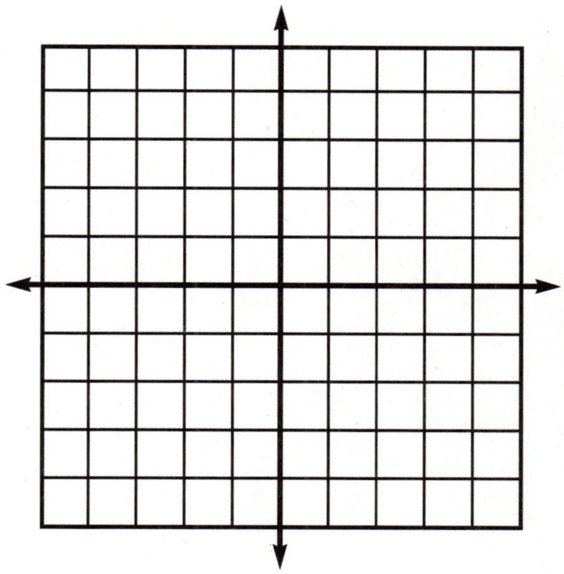

4. Describe similarities and differences between the equations and their resulting graphs. What causes this effect?

TOPIC 5

Consumer Applications

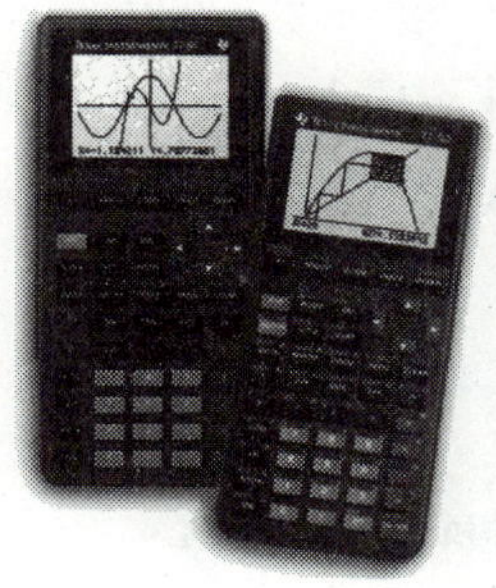

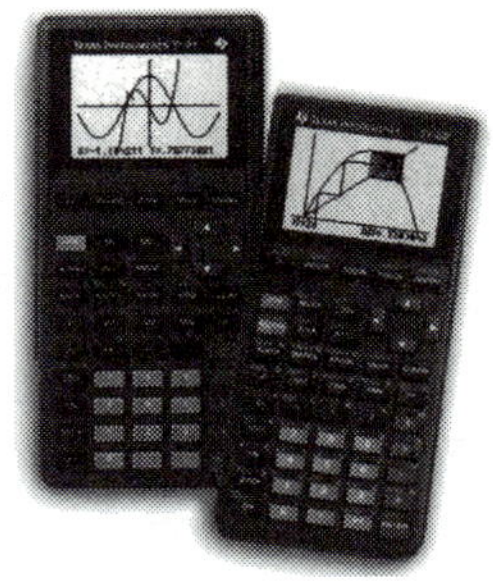

Compound Interest

Using "Compound Interest" students will

- Calculate compound interest
- Compare interest compounded annually, semiannually, and quarterly
- Choose an account based on terms and interest rates

Teaching the Lesson

- Press the MODE key to set the calculator to two decimal places.
- To calculate the new balance after one year, use $5000 \times (1+.08)$. Press ENTER to get the new balance of $5400.
- To calculate the new balance after two years, press ANS to display the previous result, then multiply by (1+.08). ENTER will get you the new balance of $5832.
- To calculate the new balance after three years, press ENTER to get a balance of $6298.36.
- To calculate the new balance after four years, press ENTER to get a balance of $6802.44.
- To calculate the new balance after five years, press ENTER to get a balance of $7346.64
- For interest compounded semiannually, use $5000(1 + \frac{0.08}{2})$ and repeat the same procedure. Repeat twice for one year's balance since each repetition is only for half a year.

After one year	$5408.00
After two years	$5849.29

- For interest compounded quarterly, use $5000(1+ \frac{0.08}{4})$ and repeat again. This time four repetitions gives one year's balance.

After one year	$5412.16
After two years	$5858.30

Compound Interest

1. Predict: In which account would you rather have your money?

   $1000 at 6% compounded annually — $1000(1 + 0.06)$ once

   $1000 at 6% compounded semiannually — $1000(1 + \frac{0.06}{2})$ twice

   $1000 at 6% compounded quarterly — $1000(1 + \frac{0.06}{4})$ four times

   Prediction ______________________________

   Now use your calculator to solve.

   Largest amount paid _____________________

   Which rate? ____________________________

   Was your prediction correct? ______________

2. Predict: In which account would you rather have your money?

   $2000 at 6% compounded annually for five years

   $2000 at 8% compounded quarterly for four years

   Prediction ______________________________

   Now use your calculator to solve.

   Largest amount paid _____________________

   Which savings plan? _____________________

   Was your prediction correct? ______________

3. Predict: In which account would you rather have your money?

   $10,000 at 12% compounded semiannually for four years

   $10,000 at 15% compounded quarterly for three years

   Prediction ______________________________

   Now use your calculator to solve.

   Largest amount paid _____________________

Which savings plan? ____________________

Was your prediction correct? ______________

4. Predict: In which account would you rather have your money?

$5000 at 10% compounded annually for two years

$5000 at 12% compounded semiannually for one year

Prediction ____________________________

Now use your calculator to solve.

Largest amount paid ____________________

Which savings plan? ____________________

Was your prediction correct? ______________

5. Predict: In which account would you rather have your money?

$3500 at 5% compounded annually for three years

$3500 at $5\frac{1}{4}$% compounded semiannually for two years

Prediction ____________________________

Now use your calculator to solve.

Largest amount paid ____________________

Which savings plan? ____________________

Was your prediction correct? ______________

6. Should a decision on a savings plan only be based on an interest rate?

7. What other factors affect your savings balance?

Profit

Using "T-Charts for T-Shirts" the students will

- Investigate a consumer problem using a chart
- Write equations for Cost, Profit, and Revenue
- Graph the profit equation for the problem and analyze the graph at a certain profit level
- Use the graphing capabilities to analyze profits

Teaching the Lesson

- Students may need some help setting up the cost, profit, and revenue equations.
- Tracing may change the range on the viewing screen.

T-Charts for T-Shirts

The student council organization decided to sell T-shirts as a fund raiser project. The cost of a solid color T-shirt from the Wee Bee Tee Shirt Co. is $5.95 per shirt. The club members wanted the mascot logo on the shirt and would have to pay an additional $2.75 per shirt. The student council will be selling the shirts for $15.00 each. How much money might the student council make?

1. Complete the following T-chart to find the different profits for the different number of shirts sold.

Number of Shirts	Cost	Revenue	Profit
0			
10			
20			
30			
40			
50			
60			
x			

Suggested Range
$X_{min} = 0$
$X_{max} = 60$
$X_{scl} = 1$
$Y_{min} = 0$
$Y_{max} = 500$
$Y_{scl} = 1$
$X_{res} = 1$

2. Why must the $X_{min} = 0$ and the $X_{max} = 60$? (What does x stand for?)

3. Why must the $Y_{min} = 0$ and the $Y_{max} = 500$? (What does y stand for?)

4. Write an equation for the cost.

5. Write an equation for the revenue.

6. Write an equation for the profit.

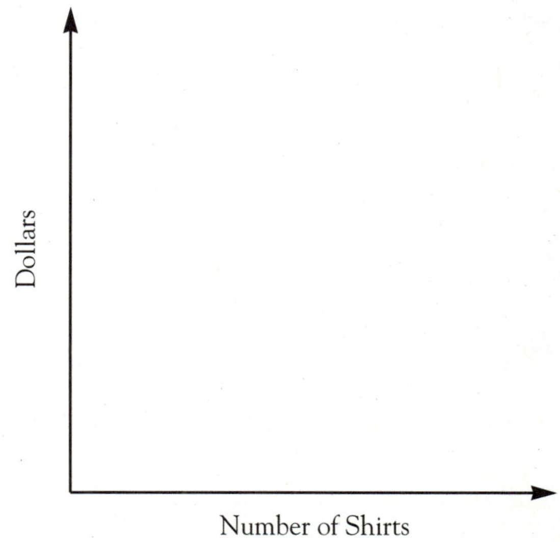

7. In the [Y=] menu, enter your profit equation for Y_1 and graph.

8. How many T-shirts must be sold to make a profit of at least $300?

9. In the [Y=] menu, enter 300 for Y_2 and graph. Use [TRACE] to find the number of T-shirts needed for a profit of $300. (Be sure you look at the correct variable!)

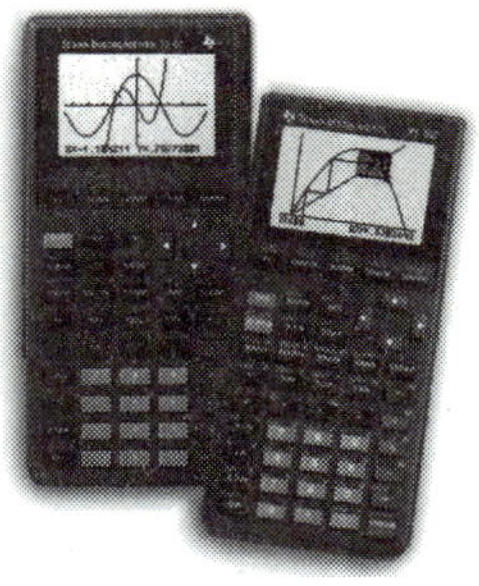

Interest Rates

Using "Buying a Set of Wheels" students will

- Compute payments given a set of time-rate values
- Use given information to make decisions regarding loan terms and rates
- Support reasons for choosing the loan and terms
- Use programming to calculate payments on a loan

Teaching the Lesson

- Students will need to review the following.

  i = periodic interest rate—the interest rate divided by the number of compounding periods

  n = the total number of payments to be made

  $$\text{Payment} = \text{Loan Value} * \frac{i * (1+i)^n}{(1+i)^n - 1}$$

- The commands for the program are located in the I/O submenu of PRGM.

Buying a Set of Wheels

Amount Financed	APR	Term of Loan
80%	8.00%	24 mos.
	8.25%	36 mos.
	8.50%	48 mos.
	8.65%	60 mos.
90%	8.40%	36 mos.
	8.65%	48 mos.
	9.00%	60 mos.
100%	9.50%	48 mos.
	10.00%	60 mos.

Use the chart above to answer the following questions.

1. Determine the interest rates available for financing

   100% of your vehicle ____________________

   90% of your vehicle ____________________

2. What happens to your interest rates when you finance 90% versus 100%?

3. If you were to finance 90% of the purchase price of your vehicle, what is the lowest interest rate available? Is there a disadvantage to selecting this interest rate?

4. What effect does a shorter term have on monthly payments?

Use the calculator to generate car payments for various interest rates and loan terms. Under the **MODE** menu, set the decimal for 2 places.

5. Enter the following program

```
:ClrHome
:Disp "INTEREST (APR)"
:Input I
:I/12→I
:Disp "NO. OF PYMNTS"
:Input N
:Disp "AMT FINANCED"
:Input P
:Disp "PAYMENT"
:P*(I*(1+I)^N)÷((1+I)^N-1)→A
:Disp A
```

6. Verify that the program is correct. Run the program for a 9.5% APR on $12,000, for 48 months. Convert percentages to decimals—for example, 9.5% = .095. You should get a monthly payment of $301.48.

7. Complete the table below for a new vehicle with a purchase price of $13,895.00.

Interset Rate		Number of Payments	Amount Financed	Payment	Total Cost of Vehicle = Payment × Number of Monthls
%	Decimal				
10%		60	$13,895.00		
9.5%		48	100% financed		
9.0%		60			
8.65%		48	$________		
8.4%		36	90% financed		
8.65%		60			
8.5%		48	$________		
8.25%		36	80% financed		
8.0%		24			

8. Why did the initial amount financed change for 90% financing and 80% financing?

9. Which car payment was the lowest? Why did this occur?

10. Which monthly car payment results in the lowest total cost for the automobile? Why?

11. Which car payment was the highest? Why?

12. Which monthly car payment results in the highest total cost for the automobile? Why?

13. List the three key factors involved in choosing a car payment and explain the effect of each.

TOPIC 6

Data Collection and Interpretation

Box and Whiskers Plot

Using "Box and Whiskers" students will

- Enter statistical data into L_1
- Determine the median and quartile values
- Use the graphing capabilities to create a box and whiskers plot
- Interpret the information represented in a box and whiskers plot

Teaching the Lesson

- This lesson uses features found only on the TI-82.
- Clear all previous drawings with ClrDraw in the DRAW menu.
- Clear the L_1 data list with ClrList in the STAT menu.
- The quartiles of a distribution are the three points that divide the set of data into four equal parts. The middle quartile is the median.
- Pay careful attention to the STAT PLOT menu directions in the lesson as they are essential to the box and whiskers plot.

Box and Whiskers

1. The ages of the players on a soccer team sponsored by the community center are listed below. Enter the data into your calculator by accessing the [STAT] menu and Edit the data into L_1.

$$\{12, 14, 8, 9, 8, 16, 12, 13, 12, 15, 13\}$$

2. Use the calculator to order the ages from youngest to oldest. To do so, use the [STAT] menu to SortA and [ENTER]. Place L_1 after the SortA command to indicate that you want to sort L_1 in ascending order and [ENTER]. To view the sorted data, select L_1 and [ENTER]. Use the right and left arrows to scroll through the data. Record the sorted data.

______, ______, ______, ______, ______, ______, ______, ______, ______, ______, ______.
min max

3. Find the lower quartile, median, and upper quartile.

LQ: ________ Med: _________ UQ: ________

4. Verify your calculations with the calculator using CALC submenu of the [STAT] menu. Select 1-Var Stats and [ENTER] twice. The quartiles and the median are labeled as Med (median), Q_1 (lower quartile), Q_3 (upper quartile). Scroll down the screen to find these values.

5. This information can be represented graphically using a box and whiskers plot. Set the proper [WINDOW] use your minimum and maximum data values for X_{min} and X_{max}. Set $Y_{min} = 0$ and $Y_{max} = 1$. To create the box and whiskers plot for this data, access the [STAT PLOT], select Plot 1. Highlight On and highlight the box and whiskers type (the third in the list), then [GRAPH].

6. Explore your box and whiskers plot using [TRACE].

7. Write a short description comparing the box and whiskers plot to the minimum, maximum, upper quartile, lower quartile, and median values of the data. Locate those values on the plot below.

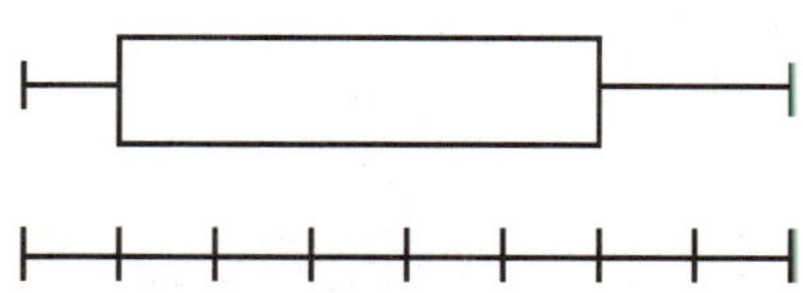

8. Answer the following questions.

a. Why is the median value not in the center of the rectangle?

b. Why is the left whisker smaller that the right whisker?

c. What does the data in the rectangle represent.

d. Draw a box and whiskers plot where there is more data closer to the median in the upper half of the data than the lower half.

9. Create a data set that meets each of the specified conditions and sketch the box and whiskers plot. Use your calculator to verify your work.

a. The box and whiskers plot has a median in the middle of the rectangle and whose whiskers are the same length.

b. The box and whiskers plot has more data closer to the median in the upper half of the data set than the lower half.

Probability: Binomial Data for Four Trials

Using "How Many Families Have Four Girls?" students will

- Sample 50 four-digit random numbers using the random number generator of the TI-81 or TI-82
- Collect data on number of girls and boys per family by using the digits of the random number as follows: girls: even digits (0, 2, 4, 6, 8); boys: odd digits (1, 3, 5, 7, 9). For example, 4702 represents 3 girls and 1 boy.
- Generate a class set of results and analyze the data on family composition

Teaching the Lesson

- Use [MODE] to set the decimal to 4 places. The random number generator is in the [MATH] menu, go to PRB, and select Rand.
- This lesson could also be explored using four dice to make comparisons of dice simulations to calculator simulations.

How Many Families Have Four Girls?

Use the random number generator of the graphing calculator to investigate the following.

Some families have four children each. How many would have all girls? Sample 50 families to find out how many would have all girls.

1. Before you begin, predict the following.

   Of the 50 families with four children, how many do you think will have

   no girls and four boys? ________________

   one girl and three boys? ________________

   two girls and two boys?________________

   three girls and one boy? ________________

   four girls and no boys? ________________

2. Under the MODE menu, select 4 decimal places.

3. Under the MATH menu, go to PRB, and select Rand.

4. Each time you press ENTER, you will get a new four-digit random number.

5. Record the number of girls (even digits) and boys (odd digits) in this simulated family of four digits.

6. Complete the chart for 50 different families.

Use the data you collected to answer the following.

7. How many families have four girls (even digits)?

8. How many families have three girls?

9. How many families have two girls?

10. How many families have one girl?

11. How many families have no girls?

12. How many families have at least three girls?

13. How many families have at most one boy?

14. What did you notice about the answers to 12 and 13?

15. What is another way of saying "How many families will have at most two girls?"

16. How do your predictions compare to the actual results?

17. What percentage of families had four girls?

18. What percentage of families had four boys?

19. What percentage of families had an equal number of boys and girls?

Family	Girls	Boys
1		
2		
3		
4		
5		
6		
7		
8		
9		
10		
11		
12		
13		
14		
15		
16		
17		
18		
19		
20		
21		
22		
23		
24		
25		

Family	Girls	Boys
26		
27		
28		
29		
30		
31		
32		
33		
34		
35		
36		
37		
38		
39		
40		
41		
42		
43		
44		
45		
46		
47		
48		
49		
50		

Probability: Experimental Versus Theoretical

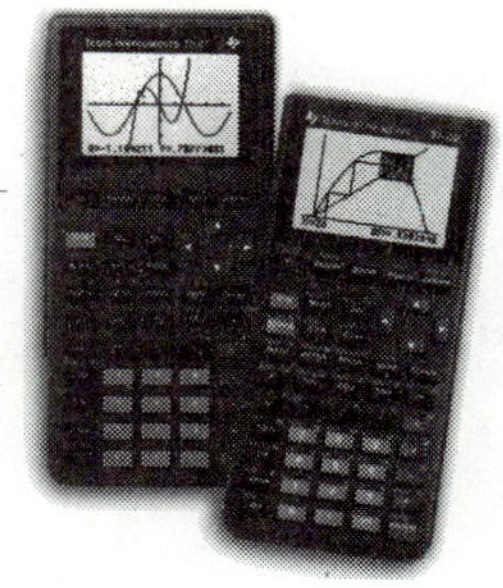

Using "Math Vegas" students will

- Input a program that simulates a random draw from a deck of cards
- Generate sample data for 50 draws and record results
- Determine probabilities based on the experiment

Teaching the Lesson

- Re-setting the seed will assure that the numbers truly are random. Students should store their phone number as the seed. To enter the seed, enter you phone number (no hyphens), press STO ▶ followed by MATH, go to PRB and select Rand, press ENTER
- The following command locations will be helpful.

Command	Where to find it.
ClrHome	PRGM menu, I/O submenu
Disp	PRGM menu, I/O submenu
to get quote marks	ALPHA, +
to get a space	ALPHA, 0
IPart	MATH menu, NUM submenu
Rand	MATH menu, PRB submenu
the little arrow	STO ▶
End	PRGM menu

Math Vegas

You are on your way to Math Vegas, Nevada. To prepare for your trip, you will use the random number generator of the TI-81 or TI-82 to simulate the drawing of cards from a standard deck.

Enter the following program.

```
Prgmx:CARDS
:ClrHome
:Disp "YOUR CARD IS"      (This line is optional.)
:IPart (Rand*13)+1→N
:IPart (Rand*4)+1→S
:Disp N
:Disp S
:End                      (for TI-82) :Stop
```

This program will generate two numbers. The first is a number between 1 and 13 and should be interpreted as follows:

1	2	3	4	5	6	7	8	9	10	11	12	13
Ace	2	3	4	5	6	7	8	9	10	Jack	Queen	King

The second number is between 1 and 4, and should be interpreted as follows:

1	2	3	4
Clubs ♣	Diamonds ♦	Hearts ♥	Spades ♠

For example, $\begin{matrix}5\\3\end{matrix}$ is the 5 of hearts. $\begin{matrix}12\\1\end{matrix}$ is the queen of clubs.

To assure that everyone gets different cards, store your phone number as the seed for the random number generator. The seed gives the random number generator a starting point. To enter the seed,

- Type in your phone number (without hyphens)
- Press [STO▶], [MATH], go to PRB and select Rand
- Press [ENTER].

1. Work with a partner to use the program to draw 50 cards. One of you should generate the cards while the other records the data and keeps count of the total number of trials. Use tally marks and the chart below to keep track of each draw.

	Clubs ♣	Diamonds ♦	Hearts ♥	Spades ♠	TOTAL
Ace					
2					
3					
4					
5					
6					
7					
8					
9					
10					
Jack					
Queen					
King					
TOTAL					

2. What was the total number of clubs drawn? __________

diamonds? __________ hearts? __________ spades? __________

3. Based on your experiment, what is the probability a club will be drawn? __________

a diamond? __________ a heart? __________ a spade? __________

4. What was the total number of aces drawn? __________

5s? __________ 10s? __________ kings? __________

5. Based on your experiment, what is the probability an ace will be drawn? __________

a 5? __________ a 10? __________ a king? __________

6. The actual probability of an event (like drawing a card) is the number of favorable outcomes (S) out of the total possible number of outcomes (T), given all outcomes are equally likely. Or,

$$P = \frac{S}{T}$$

Because there are 4 threes in the deck and a total of 52 cards, the probability of drawing a three from a standard deck is

$$P = \frac{4}{52} = \frac{1}{13} \cong 0.077 \cong 8\%$$

Using this information,

What is the probability that a club will be drawn? __________

a diamond? __________ a heart? __________ a spade? __________

What is the probability that an ace will be drawn? __________

a 5? __________ a 10? __________ a king? __________

7. How do these theoretical probabilities compare to your experimental probabilities? Explain any discrepancies?

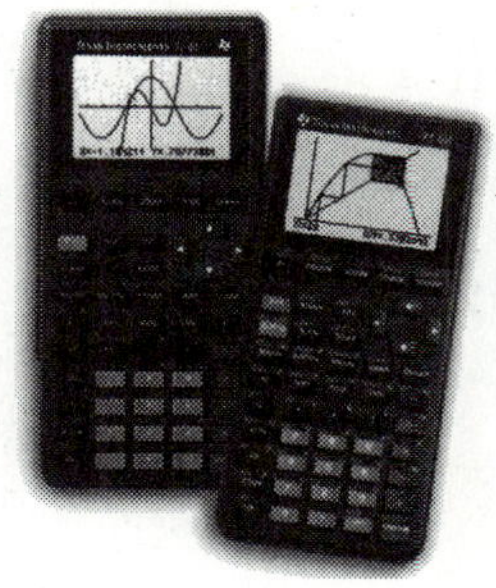

Probability: Simulation Versus Experiment

Using "Dice Toss" students will

- Generate data from an actual toss of a die
- Use the programming capabilities to run a simulation
- Generate data from a calculator simulation
- Determine probabilities of events using results from the toss of the die and the calculator simulation
- Compare and interpret results from both activities using a histogram

Teaching the Lesson

- Each group will need a die and calculator. Each group of four should divided into two pairs. The first will toss a die 10 times and record the results on the "Actual Toss Tally." The second pair will simulate 10 tosses using the "Toss the Die" program, and record the results on the "Simulated Toss Tally."
- After the first ten tosses, actual or simulated, students should exchange equipment and repeat for another set of 10 tosses. Every pair should have the opportunity to toss the die and use the calculator.
- Experimental probabilities approach theoretical probabilities more closely as the number of trials increases. Combining class data may make that more evident.

Dice Toss

1. When tossing the die, which number do you think will come up most often? Why?

2. When tossing the die, which number do you think will come up least often? Why?

3. What is the probability of rolling a four?

4. Toss the die 10–20 times. Record the results on the "Actual Toss Tally."

5. Do you think the calculator will give similar results to tossing the die?

To simulate the toss of a die on the calculator, enter the following program.

```
PrgmX:DIE
:ClrHome
:Disp "TOSS THE DIE"
:IPart (Rand*6)+1→D
:Disp D
:End
```

To assure that everyone gets different rolls of the die, store your phone number as the seed for the random number generator. The seed gives the random number generator a starting point. To change this seed,

- Type in your phone number (no hyphens)
- Press [STO▶] [ALPHA] [MATH]
- Use the left arrow to select PRB [1] to select Rand, and [ENTER].

6. Run the program and simulate the tossing of a die 10–20 times. Record the results.

7. Using data from the actual toss, what is the probability of rolling a 1? __________

2? __________ 3? __________ 4? __________ 5? __________ 6? __________

8. Using data from the calculator simulated toss, what is the probability of rolling

a 1? __________

2? __________ 3? __________ 4? __________ 5? __________ 6? __________

9. How did the results generated by the calculator compare to the results from actually tossing the die?

10. What are the advantages of using a calculator to generate data compared to actually conducting an experiment?

11. Would you change your opinion on any of the opening questions? Why or why not?

12. What would happen to the probabilities if you were to toss the die (actual and simulated) many more times? Explain.

13. How could you modify the program to simulate the tossing of a coin?

Toss Tally Sheet

Actual Toss Tally

Toss No.	1	2	3	4	5	6
1						
2						
3						
4						
5						
6						
7						
8						
9						
10						
11						
12						
13						
14						
15						
16						
17						
18						
19						
20						
Total						

Simulated Toss Tally

Toss No.	1	2	3	4	5	6
1						
2						
3						
4						
5						
6						
7						
8						
9						
10						
11						
12						
13						
14						
15						
16						
17						
18						
19						
20						
Total						

Fundamental Counting Principle

Using "The Mouse" students will

- Generate three random numbers, each between 1 and 3, simulating choices in passing through a maze
- Generate data until all possibilities are exhausted
- Be introduced to the multiplication counting principle and verify experimental and theoretical results

Teaching the Lesson

- As students generate their random numbers, it may make it easier if they clear their screen after generating the three numbers.
- The following commands can be found in:

  IPart in [MATH], NUM submenu

  Rand in [MATH], PBR submenu

- The simulations for the problems other than "The Mouse" will require simple programming similar to the example given in the student sheet.

The Mouse

A mouse is let loose in a maze. At the start of the maze, he must choose from three doors. Once he chooses one of these doors, he must again choose from three different doors, and then again, from three more doors.

1. Investigate different possible paths the mouse can take.

The simulation: Use your TI-81 or TI-82 to simulate passing through a maze. Generate a random number to represent each choice with the following command

IPart(Rand*3)+1

Press **ENTER** three times to simulate the three choices. If you get 3 1 2, for example, this means the mouse went through door 3, then door 1, then door 2.

Repeat this simulation until you think you have generated all the different possible outcomes. Keep track of your outcomes in an organized manner to make your counting easier.

2. How many possible outcomes did you get?

The number of choices can be generated by simulation, as you did, or by illustration. This is exemplified in the following drawing.

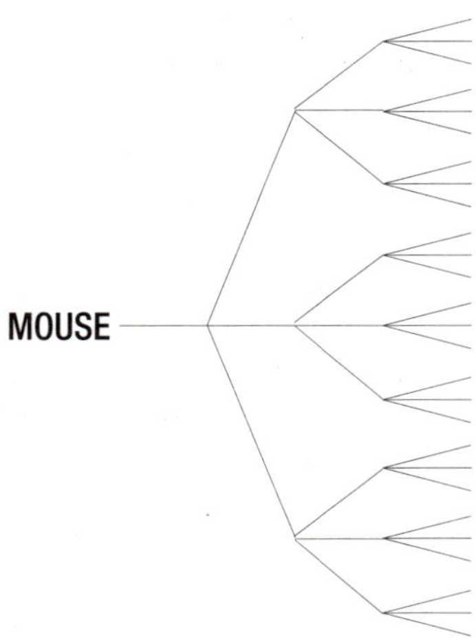

Both of these methods are cumbersome for more than a few choices. Let's try to develop a method for counting the possible outcomes.

3. How many possibilities were there for the mouse in his first choice? ______________

4. For each of those choices, how many options were available at the second decision point?

____

5. How many total choices so far? How did you arrive at this? ______________

6. For each of those choices, how many options were available at the third decision point?

____

7. How many total choices are there? ______________

8. Is this solution in line with your simulation and the drawing?

9. If you have three shirts, two pants, and four pair of shoes, how many different outfits do you have?

   a. How many shirts can you choose? ______________

   b. For each shirt, how many pair of pants can you choose? ______________

   c. How many total choices so far? ______________

   d. How did you arrive at this?

   e. For each of those choices, how many pair of shoes can you choose?

   ____

   f. How many total choices are there? ______________

   Verify this solution with a simulation or a drawing.

10. These examples illustrate the Fundamental Counting Principle. Describe the meaning of this principle.

Apply the Fundamental Counting Principle in each of the following. Then verify each by simulation or illustration.

11. How many complete meals can you make if you choose from six main dishes, five side dishes, four desserts, and three beverages?

12. A video store allows you to rent one comedy, one drama, one action/adventure film, one horror film, and one science fiction movie. How many combinations of videos can you rent if you can choose from six comedies, five dramas, four action/adventure films, three horror films, and four science fiction movies, and you rent one of each movie type?

13. You are the coach of the 1992 Olympic Dream Team. You get to choose from 40 guards, 30 forwards, and 20 centers. You need to determine the number of possible teams. Which method works best: simulation, illustration, or the counting principle? Why?

TOPIC 7

Geometry and Measurement

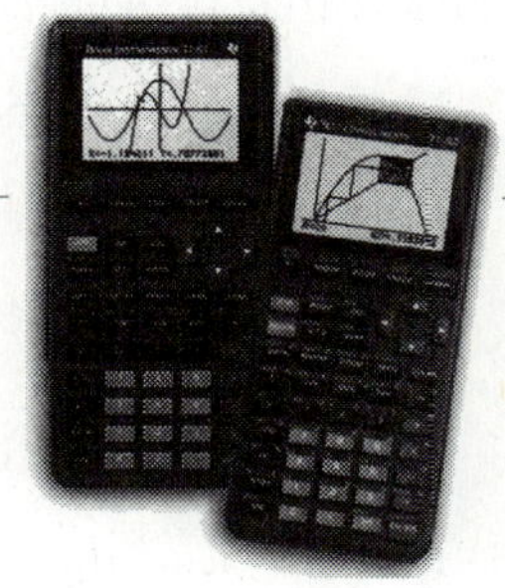

Area and Perimeter

Using "Double Trouble" students will

- Find the area and perimeter of rectangles
- Predict what happens to the area and perimeter of a rectangle when the dimensions are doubled
- Draw the rectangle using the DRAW command and compare resulting areas and perimeters
- Describe the effects of changing the dimensions
- Plot vertices of a rectangle using the plot commands on the calculator

Teaching the Lesson

- Make sure that the DRAW menu is cleared before starting the lesson.
- Students will need to set their calculator before plotting the points on the coordinate grid. If using the TI-81, the GridOn function is found in the MODE menu. The GridOn function on the TI-82 is found in the WINDOW menu.

Double Trouble

1. Under the [MODE] menu, set GridOn, if using the TI-81.

   Under the [WINDOW] menu, set GridOn, if using the TI-82.

   Suggested range

TI-81	TI-82
$X_{min} = 0$	$X_{min} = 0$
$X_{max} = 9.5$	$X_{max} = 9.4$
$X_{scl} = 1$	$X_{scl} = 1$
$Y_{min} = 0$	$Y_{min} = 0$
$Y_{max} = 6.3$	$Y_{max} = 6.2$
$Y_{scl} = 1$	$Y_{scl} = 1$

   IMPORTANT: Press [GRAPH].

2. Plot the rectangle whose vertices are the points (5, 3), (5, 2), (3, 2), (3, 3)

   Use the [DRAW] menu to select the Line option.

   Use your arrow keys to place your cursor on (5, 3) [ENTER]

   Use your arrow keys to place your cursor on (5, 2) [ENTER] [ENTER]

   Use your arrow keys to place your cursor on (3, 2) [ENTER] [ENTER]

   Use your arrow keys to place your cursor on (3, 3) [ENTER] [ENTER]

   Use your arrow keys to place your cursor on (5, 3) [ENTER]

3. What is the area of the rectangle?

4. What is the perimeter of the rectangle?

5. Starting at (5, 3), turn your draw command on with [ENTER]. Use the arrow keys to double the length of the rectangle. What are the coordinates of this new endpoint?

6. Use the arrow keys to double the width of the rectangle. What are the coordinates of this new endpoint?

7. Complete the rectangle. What are the coordinates of the vertices of the rectangles?

Small Rectangle	Large Rectangle
(_____, _____)	(_____, _____)
(_____, _____)	(_____, _____)
(_____, _____)	(_____, _____)
(_____, _____)	(_____, _____)

8. What are the dimensions of each rectangle?

Small Rectangle		Large Rectangle	
length	________	length	________
width	________	width	________

9. Use the coordinates to explain the dimensions.

10. What is the area of the small rectangle? What is the area of the large rectangle?

11. How do these compare?

12. What is the perimeter of the small rectangle? What is the perimeter of the large rectangle?

13. How do these compare?

14. Continue the investigation by clearing the draw screen. Plot the rectangle with vertices at (1, 1), (1, 3), (4, 3), (4, 1). Double the dimensions and compare the areas and perimeters of the two rectangles.

15. How does doubling the length and width of a rectangle affect the area and perimeter? Try other rectangles if necessary.

16. What do you expect happens if you triple the dimensions? Test your conjecture. Can you generalize what happens if you take n times a dimension?

17. Use the area and perimeter formulas to show the effects of doubling dimensions and compare to your results. $A = lw$, $P = 2(l + w)$.

Area of a Circle

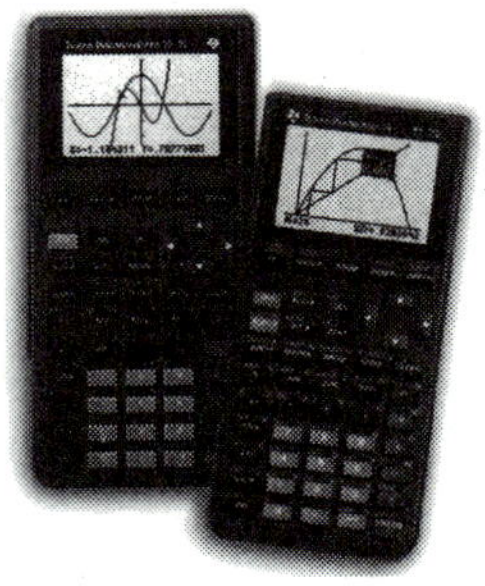

Using "Pump Up the Area" students will

- Collect data and organize it in a chart
- Determine how a scale factor change in the radius of a circle changes its area
- Input and graph the volume formula for a circle
- Use the edit features of the calculator to calculate area of a circle
- Compare area ratios

Teaching the Lesson

- Substituting in the values for doubling the radius and then entering them into the chart before simplifying, may help students recognize the pattern. The first chart focuses on doubling the radius while the second chart triples the radius.
- Students may need to create their own chart when answering number 5b.

Pump Up the Area

1. Enter $Y_1 = \pi x^2$.

2. GRAPH the area using the suggested range

$X_{min} = -1$
$X_{max} = 94$
$X_{scl} = 2$
$Y_{min} = -1$
$Y_{max} = 1259$
$Y_{scl} = 10$
$X_{res} = 1$

3. Complete the following using the TRACE function.

		Formula	Area	Ratio $\frac{A_2}{A_1}$
r	2	$\pi(2)^2$	$A_1 =$	
2r	4	$\pi(4)^2$	$A_2 =$	
r	3		$A_1 =$	
2r			$A_2 =$	
r	5		$A_1 =$	
2r			$A_2 =$	

a. What do you notice about the ratios?

b. When the radius of a circle is doubled, the area increases by a factor of ______.

4. Complete the following using the TRACE function.

		Formula	Area	Ratio $\frac{A_2}{A_1}$
r	2	$\pi(2)^2$	$A_1 =$	
$3r$			$A_2 =$	
r	3		$A_1 =$	
$3r$			$A_2 =$	
r	5		$A_1 =$	
$3r$			$A_2 =$	

a. What do you notice about the ratios?

b. When the radius of a circle is tripled, the area increases by a factor of _______.

5. Generalize.

a. If we quadruple the radius of a circle, the area increases by a factor of _______. Verify your answer.

b. If we take half the radius of a circle, the area increases/decreases by a factor of _______. Verify your answer.

Radius and Volume of a Sphere

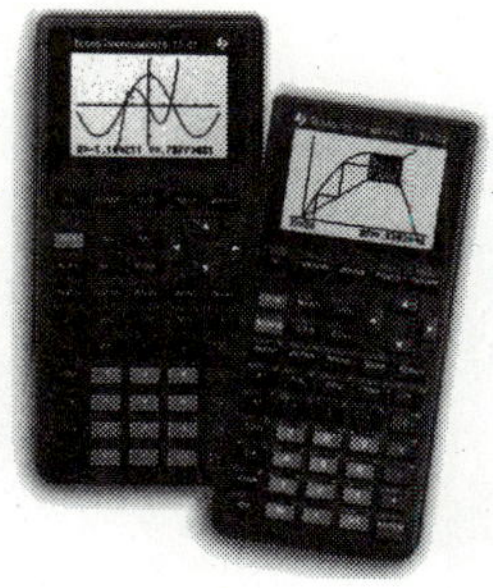

Using "Pump Up the Volume" students will

- Calculate the volume of a sphere given the radius
- Use the graphing capabilities to graph the volume using values from the chart
- Use the graph to make predictions
- Make predictions for volumes given radii
- Describe in writing the effect of a change in the radius on the volume of the sphere

Teaching the Lesson

- This activity is designed to be used with small groups or pairs.
- Each group will need a ruler and five round balloons. It may be helpful to have three 9-inch diameter balloons and two 12-inch diameter balloons. When measuring the diameters of the balloons, draw the desired diameter on a sheet of paper and inflate the balloon until it reaches the desired diameter.
- The students' predictions will improve as they work through the five examples with the balloons.
- Make sure that the [Y=] menu is cleared before beginning the lesson.

Pump Up the Volume

1. Inflate each of the 5 balloons to the various sizes shown in the chart below. Enter the color of each balloon and its diameter in the chart. Since most balloons are not exactly spherical, our balloons will be models for spheres approximated by the balloons. Round all answers to the nearest hundredth. Use the edit command to save time. Use your previous answer to help you make more accurate predictions as you complete the chart.

	Diameter of Balloon	Radius of Balloon	Predicted Volume of Balloon	Volume of the Sphere $V = \frac{4}{3}\pi r^3$	Approximate Volume of Balloon
balloon 1		2 in.		$\frac{4}{3}\pi(\_)^3$	
balloon 2		3 in.		$\frac{4}{3}\pi(\_)^3$	
balloon 3		4 in.		$\frac{4}{3}\pi(\_)^3$	
balloon 4		5 in.		$\frac{4}{3}\pi(\_)^3$	
balloon 5		6 in.		$\frac{4}{3}\pi(\_)^3$	

2. To graph the volume, use the suggested range

$X_{min} = 0$
$X_{max} = 19$
$X_{scl} = 1$
$Y_{min} = -504$
$Y_{max} = 4410$
$Y_{scl} = 200$
$X_{res} = 1$

3. Enter the volume formula for Y_1. Graph the volume. Sketch the graph.

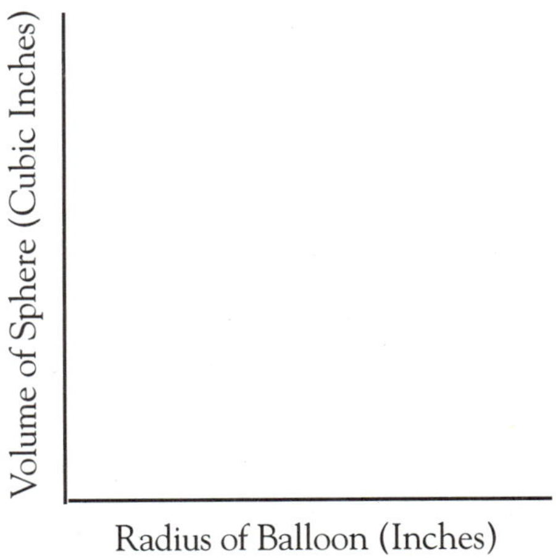

4. Use the TRACE feature to find the volume of spheres with radii 2, 3, 4, 5, and 6 inches. Record your results. Do these values correspond to the values in the chart in question 1?

Radius of Sphere	Volume of Sphere
2 inches	
3 inches	
4 inches	
5 inches	
6 inches	

5. Use the TRACE feature to approximate the volume of balloons with radii of 8 inches, 9 inches, and 10 inches.

6. In a brief paragraph, explain how the volume of the sphere changes as the radius changes.

7. What is the approximate volume of a sphere with radius of –3 inches? Explain your answer.

8. Do you think the volume of a sphere with a radius of 6 centimeters would be greater or less than a balloon with a radius of 6 inches? Explain.

TOPIC 8

Programs for the TI-81 and TI-82

Sample Programs

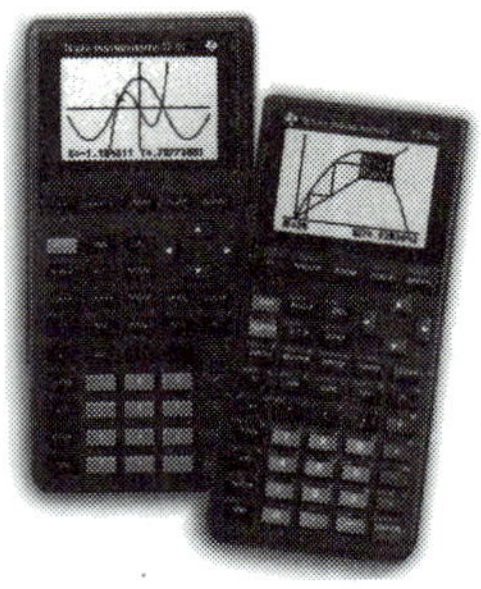

Using these "Sample Programs" students can

- See how everyday, tedious mathematical operations can be programmed
- Analyze the purposes of each step in a program
- Test the programs
- Evaluate programs for efficient use of available memory
- Modify the programs
- Use the programs as examples as they translate other mathematical jobs into machine commands

Teaching with the Sample Programs

- The language used by the graphing calculator has some limitations, but the ease of working with commands makes it possible to program using minimum time on computer commands and maximum time on the mathematical process.
- Suggestions for modifying some of the programs are included.
- Encourage students to modify and test modifications for any program as they experiment with programming. They can use the sample programs as models. For example, they can apply formula programs to new formulas.
- Discuss with your students the balance between efficient use of memory and understandability of the program. For example, two versions of the sales tax program are suggested. The shorter is more efficient in memory use, but someone seeing that program for the first time may not have clear directions on its use. The second program uses more memory but is easier to use.
- Programs such as "Circle Vocabulary" allow students to review the terms used for circles as they enter and use the program. They can then use the program to teach these terms to another student.
- Students can use programs to answer a variety of questions. For example, "Dinner Cost" can be used to calculate the total cost of a dinner. But students can also be challenged to determine what they may spend for dinner given a fixed limit on the total.
- If students use their own machine, they can store their programs. Students who share calculators may find some programs already entered and learn from other student's work.
- Some of the sample programs are given names. You can change these names as needed to work with other programs saved on your graphing calculator.

- For some of the sample programs, the program for the TI-81 and TI-82 are similar; for others, the programs do different things.
- Subroutines on the TI-81 are referenced by number, Prgm8. Subroutines on the TI-82 are referenced by name.
- There are some differences between programming the TI-81 and the TI-82. On the TI-82

    END has been replaced with STOP because END statements have a new purpose.

    It is necessary to rewrite comments on Disp statements to 16 characters or less.

    When you are making changes to the range values, press [WINDOW].

    When you press [STO▶] you must press [ALPHA] to get into alpha mode.

    Programs can be directly transferred from one TI-82 to another.

Sample Programs

Sales Tax

These programs for computing sales tax illustrate that different commands can accomplish the same job. Think about which program uses memory efficiently and which is easiest for many people to use.

TI-81

```
Prgm:TAX
:Fix 2
:Disp "COST"
:Input C
:Disp "TAX RATE"
:Input T
:C*T→S
:Disp S
:C+S→P
:Disp P
```

TI-82

```
PROGRAM:SALESTX1
:Fix 2:Disp "COST":Input C
:Disp "TAX RATE"
:Input T
:C*T→S:Disp S
:C+S→P:Disp P
```

TI-81

```
:Fix 2
:Disp "ENTER SALES TAX RATE AS A
   PERCENT"
:Input R
:R/100→R
:Lbl 1
:Disp "ENTER THE ORIGINAL COST"
:Input C
:R*C→T
:C+T→P
:ClrHome
:Disp "COST"
:Disp C
:Disp "SALES TAX"
:Disp T
:Disp "TOTAL PRICE"
:Disp P
:Pause
:Goto 1
:End
```

TI-82

```
PROGRAM:SALESTX2
:Fix 2:Disp "ENTER SALES TAX"
:Disp "RATE AS A PERCENT"
:Input R
:R/100→R:Lbl 1
:Disp "ORIGINAL COST"
:Input C
:C+T→P
:ClrHome
:Disp "COST"
:Disp C
:Disp "SALES TAX"
:Disp T
:Disp "TOTAL PRICE"
:Disp P
:Pause
:Goto 1
:End
```

Cost of a Call

These programs will calculate the cost of a telephone call including base charge and per-minute charge. Expand the program to include variables such as call waiting or call return. Imitate these programs for other budgeting computations based on functions.

TI-81

```
Prgm: PHONE
:ClrHome
:Fix 2
:Disp "MINUTES"
:Input M
:Disp "COST OF 3 MINUTES"
:Input T
:Disp "COST PER MINUTE AFTER THAT"
:Input A
:(M-3)A+T→F
:ClrHome
:Disp "TOTAL COST IS"
:Disp F
:End
```

TI-82

```
PROGRAM:PHONE
:ClrHome
:Fix 2
:Disp "MINUTES"
:Input M
:Disp "COST OF 3 MIN"
:Input T
:Disp "COST ADDED MIN"
:Input A
:(M-3)A+T→F
:ClrHome
:Disp "TOTAL COST IS"
:Disp F
:Stop
```

Patterns from Equations

Using the TI-81, enter a number, and the computer randomly chooses one of three expressions and uses the entered number to evaluate that expression. Enter another number, and look for the pattern in the computers response. If you do know the pattern, enter the pattern as an expression such as $x - 56$. The main program, Prgm1, includes the expressions. Prgm2 is the subroutine used to enter you test values. Prgm3 is the subroutine used to ask if you recognize the pattern and to test your pattern choices to see if they are correct.

The program for the TI-82 uses the entered number as the input for the expression $4x + 5$. The program requests a group of four domain values and then displays the domain values and their respective range values in sets. When you are ready to explain the function rule, enter a number and the operation following the order of operations. (Type the number, press [ENTER], then select the operation.) Repeat if the pattern is not yet complete. (For this function, type 4, [ENTER], ×; 5, [ENTER], +.) Modify the line 4*L1+5→L2, and challenge other students to determine the new function pattern.

TI-81

```
Prgm1
:Lbl 0
:Prgm 2
:X+9→W
:Disp Ans
:Prgm 3
:If A ≠ 1
:Goto 0
:Pause
:Lbl 2
:Prgm 2
:X-5→W
:Disp Ans
:Prgm 3
:If A ≠1
:Goto 2
:Pause
:Lbl Y
:Prgm 2
:2X+1→W
:Disp Ans
:Prgm 3
:If A≠1
:Goto Y
:Stop

Prgm2
:Disp "ENTER A VALUE"
:Input X
:End

Prgm3
:Disp "DO YOU SEE THE PATTERN? 1 FOR
   YES, 2 FOR NO"
:Input A
:If A≠1
:End
:Disp "ENTER THE PATTERN"
:Input Y
:If Y=W
:Goto R
:Disp "NO, TRY AGAIN"
:2→A
:Lbl R
:If Y=W
:Disp "CONGRATULATIONS"
:End
```

TI-82

```
PROGRAM:FUNCPAT
:Lbl 0
:Input A
:A→L1(1)
:Input B
:B→L2(2)
:Input C
:C→L3(3)
:Input D
:D→L4(4)
:L1→L4
:4*L1+5→L2
:Disp L1,L2
:Disp "NUMBER"
:Lbl 3
:Input E
:Menu("OPERATION","ADD",F,"SUB-
   TRACT",G,"MULTIPLY",H,"DIVIDE",I
   )
:Lbl F:L1+E→L3:Goto 2
:Lbl G:L1-E→L3:Goto 2
:Lbl H:L1*E→L3:Goto 2
:Lbl I:L1/E→L3:Goto 2
:Lbl 2
:Menu("PATTERN COMPLETE",
   "NO",K,"YES",L)
:Lbl K:L3→L1:Goto 3
:Lbl L
:Disp L4,L2,L3
:L2-L3→L5
:If L5(1)=0
:Then
:Disp "YOU ARE RIGHT"
:Goto 5 :End
:Disp "TRY AGAIN"
:Goto 0
:Lbl 5
:Stop
```

Dinner Cost

These programs are used to calculate the tax, tip, and total cost of a restaurant dinner. The tax rate and tipping percent are entered as percents, not as decimals. Similar programs can be written to calculate such things as sales tax, shipping costs, or special handling charges on orders.

TI-81

```
:ClrHome
:Disp "TAX RATE"
:Input R
:Disp "TIPPING PERCENT"
:Input Q
:Disp "ORIGINAL COST"
:Input C
:CR/100→T
:CQ/100→G
:C+T+G→S
:ClrHome
:Fix 2
:Disp C
:Disp "TAX"
:Disp T
:Disp "TIP"
:Disp G
:Disp "TOTAL COST"
:Disp S
:End
```

TI-82

```
PROGRAM:DINNER
:ClrHome
:Disp "TAXRATE PERCENT"
:Input R
:Disp "TIPPING PERCENT"
:Input Q
:Input "ORIGINAL COST", C
:CR/100→T
:CQ/100→G
:C+T+G→S
:ClrHome
:Fix 2
:Disp C
:Disp "TAX", T
:Disp "TIP", G
:Disp "TOTAL COST", S
:Pause
:Stop
```

Arithmetic Mean

These programs compute the average of a set of numbers. For the TI-81 program you specify the number of entries to be averaged, and the mean of these numbers is displayed. The program for the TI-82 will count and display the number of grades, the individual grades, and the running average, so you can see the effect of each grade. Each time the screen halts, you can press [ENTER] to resume display and press it again to receive the next value. When the averages line gets too long to view, press ▲, up cursor, and ▶, right cursor, to scroll to the rest of the averages. Press [ON] 2 to break the program when you are finished entering grades, then press [STAT] 1 to view the table of scores and averages.

TI-81

```
:Disp "HOW MANY NUMBERS?"
:Input N
:1→C
:0→S
:Disp "DATA"
:Lbl 1
:Input X
:S+X→S
:If S>(C,N)
:Goto 1
:S/N→M
:Disp "MEAN"
:Disp M
:End
```

TI-82

```
PROGRAM:AVERAGE
:ClrList L1,L2,L3: Fix 0
:0→T
:0→C
:ClrHome
:Lbl 1
:Disp "ENTER VALUE"
:Input V
:If V>100: Then: Goto 2: End
:C+1→C
:C→L1(C)
:C→L2(C)
:T+V→T
:T/C→A
:Disp C, "ITEMS"
:Disp A, "PRESS ENTER"
:Pause
:A→L3(C)
:Disp "NUM GRADES",L1
:Disp "UP/RIGHT SCROLL"
:Disp "AVERAGE"
:Pause L3
:Goto 1
:Lbl 2
:Stop
```

The 21 Game

Two players, alternating turns, select a digit to add to the cumulative total. To win, reach a total of 21 on your turn and prevent your opponent from reaching it first. The program for the TI-81 allows you to choose the digits 1, 2, or 3. The program for the TI-82 utilizes the CUBE program (TI-82 version). Be sure it has been entered before operating this program. To play the 21 game on the TI-82, you have a choice of four of the digits between 1 and 6 on each move. After the first move, a random roll of the die, each player turns the cube to an adjacent face and adds that number to the cumulative total. Remember the typical number cube (die) is arranged so that opposite sides total 7. Therefore only certain numbers are available as each turn is made. For example, if the one is on the top face, you can only turn the cube to 2, 3, 4, or 5 because the 6 is on the bottom face. Like the game using the TI-81 program, you want to reach a total of 21 on your turn and prevent your opponent from reaching it first. These program may be modified to play the game with different totals or different operations.

TI-81

```
:Fix 0
:Disp "FIRST TO REACH 21 WINS"
:0→T
:Lbl S
:Disp "INPUT DIGIT FROM 1-3"
:Input N
:N+T→T
:Disp "CURRENT TOTAL"
:Disp T
:If T=21
:Goto E
:Goto S
:Lbl E
:Disp "YOU WON"
```

TI-82

```
PROGRAM:GAME21
:Split
:prgmCUBE
:N→T
Lbl 1
:Disp "CHOOSE ONE"
:If N=1 or N=6
:Disp {2,3,4,5}
:If N=2 or N=5:
:Disp {1,3,4,6}
:If N=3 or N=4
:Disp {1,2,5,6}
:Input P
:FullScreen
:T+P→T:Disp "TOTAL IS", T
:If T>21:Then
:Disp "TOO BIG YOU LOSE":Goto 2:End
:If T=21
:Then
:Disp "YOU WIN"
:Else
:Disp "NEXT PLAYER"
:P→N
:Disp "NUMBER ON TOP", N
:Goto 1
:Lbl 2
:Stop
```

Adjusting Recipes

These programs for the TI-81 allow you to adjust recipe measurements. Programs 7 and 8 are subroutine programs that will be called for by the main program.

TI-81

```
:ClrHome
:Disp "ADJUST RECIPE, 1 TO INCREASE, 2 TO DECREASE"
:Input D
:ClrHome
:Disp "ENTER THE NUMBER YOU WANT TO MULTIPLY OR DIVIDE BY"
:Input C
:ClrHome
:Lbl B
:If D=1
:Prgm 7
:If D=2
:Prgm 8
:Goto B

Prgm7:MULTIPLY
:Disp "ENTER ORIGINAL MEASUREMENT"
:Input Q
:CQ
:Disp "ADJUSTED"
:Disp Ans

Prgm8:DIVIDE
:Disp "ENTER ORIGINAL MEASUREMENT"
:Input Q
:Q/C
:Disp "ADJUSTED"
:Fix 2
:Disp Ans
```

Measurement Conversions

These programs will help you convert measurements. You enter the original measurement, identify the necessary operation, and supply the proper conversion factor (for the TI-81). On the TI-82, you select from a menu, and the conversion factor is supplied for you. These programs can be adjusted to convert other measurements.

TI-81

```
:ClrHome
:Disp "ORIGINAL MEASUREMENT"
:Input U
:Disp "TYPE"
:Disp "CONVERSION"
:Disp "FACTOR"
:Input C
:Disp "ENTER 1 FOR LARGE TO SMALL"
:Disp "ENTER 2 FOR SMALL TO LARGE"
:Input X
:If X=1
:U*C→A
:If X=2
:U/C→A
:Disp "NEW UNIT"
:Disp A
```

TI-82

```
PROGRAM:CONVERT
:ClrHome
:Input "VALUE TO CONVERT",M
:Menu("CONVERTING", "LENGTH", A,
   "AREA", B, "VOLUME", C)
:Lbl A
:Menu("LENGTH", "FT TO IN", D, "IN
   TO FT", E, "FT TO YD", F, "YD TO
   FT", G, "YD TO IN", H, "IN TO
   YD", I)
:Lbl D
:M*12→T:Goto 2
:Lbl E
:M/12→T:Goto 2
:Lbl F
:M/3→T:Goto 2
:Lbl G
:M*3→T:Goto 2
:Lbl H
:M*36→T:Goto 2
:Lbl I
:M/36→T:Goto 2
:Lbl B
:Menu("AREA", "SQ IN TO SQ FT", J,
   "SQ FT TO SQ IN", K, "SQ FT TO
   SQ YD", L, "SQ YD TO SQ FT", N,
   "SQ MI TO ACRES", O , "ACRES TO
   SQ MI", P)
:Lbl J
:M/144→T:Goto 2
:Lbl K
:M*144→T:Goto 2
:Lbl L
:M/9→T:Goto 2
```

TI-82 (continued)

```
:Lbl N
:M*9→T:Goto 2
:Lbl O
:M*640→T:Goto 2
:Lbl P
:Menu("VOLUME", "CU FT TO CU IN", Q,
   "CU IN TO CU FT", R, "CU YD TO
   CU FT", S, "CU FT TO CU YD", U)
:Lbl Q
:M*1728→T:Goto 2
:Lbl R
:M/1728→T:Goto 2
:Lbl S
:M*27→T:Goto 2
:Lbl U
:M/27→T:Goto 2
:Lbl 2
:Disp T
:Stop
```

Greatest Common Factor

These program will compute the greatest common factor for two numbers. Modify them to accept a fraction and express it in lowest terms. The program for the TI-82 will also compute the Least Common Multiple for a pair of positive integers.

TI-81

```
:Disp "INTEGER 1"
:Input A
:Disp "INTEGER 2"
:Input B
:Lbl R
:If A=B
:Goto 5
:If A<B
:Goto 4
:A-B→A
:Goto R
:Lbl 4
:B-A→B
:Goto R
:Lbl 5
:Disp "GCF IS"
:Disp A
```

TI-82

```
PROGRAM:GCFLCM
:Input "INTEGER",A
:Input "INTEGER", B
:A→E
:B→F
:Lbl R
:If A=B
:Goto 5
:If A< B
:Goto 4
:A-B→A
:Goto R
:Lbl 4
:B-A→B
:Goto R
:Lbl 5
:Disp "GCF IS", A
:Disp "LCM IS", E*F/A
```

Circle Vocabulary

This program for the TI-82 illustrates graphics demonstrations you can create. Creating the graphics, you will work with coordinate graphing and text. Press ENTER to see each vocabulary term displayed.

TI-82

```
PROGRAM:CIRCLE
:Split
:ClrHome
:-4→Xmin
:4→Xmax
:0→Ymin
:4→Ymax
:ZSquare
:Circle(1.5,1.5,1)
:Disp "CIRCLE"
:Pause:ClrHome
:Pt-On(1.5,1.5)
:Disp "CENTER"
:Pause:ClrHome
:Line(1.5,1.5,1.5,2.5)
:Disp "RADIUS"
:Pause:ClrHome
:ClrDraw
:Circle(2,2,2)
:Line(0,2,4,2)
:Disp "DIAMETER"
```

Angles of a Triangle

These programs will compute the measure of the third angle in the triangle given the first two angles. Adapt the program for quadrilaterals or other polygons.

TI-81

```
:Disp "FIRST ANGLE"
:Input F
:Disp "SECOND ANGLE"
:Input S
:180-(F+S)→T
:Disp "THIRD ANGLE"
:If T≤0
:Disp "NONE"
:If T > 0
:Disp T
:Pause
:End
```

TI-82

```
PROGRAM:ANGLES
:Input "FIRST ANGLE", F
:Input  "SECOND ANGLE", S
:180-(F+S)→T
:Disp "THIRD ANGLE"
:If T≤ 0
:Then
:Disp "None"
:Else
:Disp T
:Stop
```

Timer

These programs will serve as timers and can become part of a program or stored as a separate program and accessed by the subroutine command. Changing the value 200 to a different number, changes the pause time.

The TI-82 program can be used to control the display of other programs. It will allow you to study the capabilities and speed of computers, and to explore number patterns. Modify the amount or increments, to see different patterns. The OUTPUT command can be used to plan the display. This can also be modified.

TI-81

```
:0→T
:Lbl 1
:T+1→T
:If T=200
:Goto E
:Goto 1
:Lbl E
:End
```

TI-82

```
PROGRAM:TIMER
:Fix 0
:0→T
:For(T,0,200,1)
:Output(1,13,T)
:Stop
```

Time Conversion

This program was designed for the TI-81 to convert time based on a 12-hour clock to military time based on 24 hours.

TI-81

```
:Lbl Q
:ClrHome
:Fix 2
:Disp "WHAT TIME IS IT?"
:DIsp "ENTER WITH . INSTEAD OF :"
:Input T
:Disp "IS IT AM(1) OR PM(2)"
:Input S
:If S=2
:T+12→M
:If S=1
:T→M
:If Int M=12
:T-12→M
:If Int M=24
:M-12→M
:Disp M
:Pause
:Goto Q
:End
```

Number Cube

This set of programs will display a randomly chosen face of a number cube. The main program, RNG, accesses the subroutine programs A–E. Adjust program numbers as appropriate for your calculator.

TI-81

```
Prgm:RNG
:1→Xscl
:1→Yscl
:0→Xmin
:0→Ymin
:10→Ymax
:15→Xmax
:ClrDraw
:Disp "TO THROW THE NUMBER CUBE,
   PRESS ENTER"
:Pause
:Int (Rand6)+1→N
:Prgm E
:If N=2
:Prgm C
:If N=3
:Prgm A
:If N=4
:Prgm D
:If N=5
:Prgm A
:If N=5
:Prgm B
:If N=6
:Prgm C
:If N=6
:Prgm D

PrgmE:CUBE
:Line(2,2,4,2)
:Line(4,2,4,4)
:Line(4,4,2,4)
:Line(2,4,2,2)
:PT-On(3,3)
:End
```

TI-82

```
PROGRAM:CUBE
:1→Xscl
:1→Yscl
:1→Xmin
:1→Ymin
:4→Ymax
:8→Xmax
:Split
:Lbl 8
:Disp "THROW CUBE"
:Disp "PRESS ENTER"
:Pause
:ClrDraw
:int(rand6)+1→N
:Line(2,2,4,2)
:Line(4,2,4,4)
:Line(4,4,2,4)
:Line(2,4,2,2,)
:If N=1
:Then
:Pt-On(3,3)
:End
:If N=2
:Then
:Pt-On(2.5,3)
:Pt-On(3.5,3)
:End
:If N=3
:Then
:Pt-On(3,3)
:Pt-On(2.5,2.5)
:Pt-On(3.5,3.5)
:End
:If N=4
:Then
:Pt-On(2.5,2.5)
```

TI-81 (continued)

```
PrgmA:C3
:PT-On(2.5,2.5)
:PT-On(3.5,3.5)
:End

PrgmB:C5
:PT-On(2.5,3.5)
:PT-On(3.5,2.5)
:End

PrgmC:C2
:PT-Off(3,3)
:PT-On(2.5,3)
:PT-On(3.5,3)
:End

PrgmD:C4
:PT-Off(3,3)
:PT-On(2.5,2.5)
:PT-On(2.5,3.5)
:PT-On(3.5,2.5)
:PT-On(3.5,3.5)
:End
```

TI-82 (continued)

```
:Pt-On(3.5,2.5)
:Pt-On(3.5,3.5)
:Pt-On(2.5,3.5)
:End
:If N=5
:Then
:Pt-On(2.5,2.5)
:Pt-On(3.5,2.5)
:Pt-On(3.5,3.5)
:Pt-On(2.5,3.5)
:Pt-On(3,3)
:End
:If N=6
:Then
:Pt-On(2.5,3)
:Pt-On(3.5,3)
:Pt-On(2.5,2.5)
:Pt-On(2.5,3.5)
:Pt-On(3.5,2.5)
:Pt-On(3.5,3.5)
:End
```

Guessing Game

Play this game to guess the number chosen at random by the computer.

TI-81

```
:ClrHome
:Int(Rand*100)→A
:0→X
:Fix 0
:Disp "PICK A NUMBER FROM 0-100"
:Lbl 1
:X+1→X
:Input N
:If N=A
:Disp "YOU GOT IT"
:If N>A
:Disp "LOWER"
:If N<A
:Disp "HIGHER"
:If N=A
:Goto 2
:Goto 1
:Lbl 2
:Disp X
:Disp "GUESSES"
:End
```

Multiplication Flash Cards

The program, for the TI-81, will generate multiplication problems. Answer each problem, and check the answer. The TI-82 programs asks you to estimate the answer to a multiplication problem within 5%.

TI-81

```
:ClrHome
:Rand*10→X
:Round (X,0)→X
:Rand*10->Y
:Round(Y,0)→Y
:Fix 0
:ClrHome
:Lbl 1
:Disp X
:Disp "X"(go to 2nd line and four
   more spaces)
:Disp Y
:Input A
:XY→B
:If A=B
:Disp "RIGHT"
:If A≠B
:Disp "TRY AGAIN"
:If A≠B
:Goto 1
:End
```

TI-82

```
PROGRAM:ESTIMATE
:Fix 0
:0→E
:0→C
:0→P
:Lbl 1
:ClrHome
:rand10000→A
:rand 100→B
:Output(8,1,"ESTIMATE ANSWER")
:Lbl 5
:Output(5,8,A)
:Output(6,5,"X")
:Output(6,10,B)
:P+1->P
:Input C
:If abs(C-AB)<.05AB
:Then
:Disp "GOOD ESTIMATE"
:E+1->E
:ClrHome
:Disp "YOU ARE ESTIMATING"
:Disp "WITHIN 5 PERCENT",(E/P)*100
:Disp "PERCENT OF"
:Disp "THE TIME"
:For(T,0,500,1)
:T+1→T
:End
:Else
:ClrHome
:Output(2,1,"FIX ESTIMATE")
:Goto 5
:End
:Disp "READY TO DO ANOTHER PROBLEM?"
:Menu("ANOTHER PROBLEM?","YES",V, "NO",
   W)
:Lbl V
:Goto 1
:Lbl W
:ClrHome
:Disp "THANKS FOR"
:Disp "ESTIMATING"
:Stop
```

Root Estimates

This program will compute the square root of a number using the divide and average method. You estimate the first square root. For the TI-82, change the last line in the program to Stop.

TI-81 and TI-82

```
Prgm:SQROOT
:Int(Rand*100)→X
:Disp Ans
:Disp "GUESS THE SQUARE ROOT"
:Input A
:Lbl 1
:ClrHome
:X/A→B
:Disp B
:Disp A
:Pause
:100000→J
:If Int JB=Int JA
:Goto W
:(B+A)/2→A
:Goto 1
:Disp A
:Lbl W
:ClrHome
:Disp B
:Disp A
:Disp "SQROOT OF"
:Disp X
:End
```

Survey Tallies

The following program asks a survey question, tallies the results, and prints the group record. For the TI-81, the group record is displayed when the special code, 23776, is entered. Answers other than the specified choices are ignored. The variables *L* and *D* can be accessed to keep track of the tallies of several groups. The question can be designed for any survey. The TI-82 offers you answer choices and reports the group results when you press 3.

TI-81

```
:0→Y
:0→N
:0→S
:Lbl 1
:0→C
:Lbl 2
:C+1→C
:S+1→S
:Lbl 3
:ClrHome
:Disp "HAVE YOU
LIVED A MILLION
SECONDS"
:Disp "PRESS"
:Disp "1 FOR YES"
:Disp "2 FOR NO"
:Input A
:If A=23776
:Goto D
:If A=1
:Goto Y
:If A=2
:Goto N
:Disp "1 OR 2"
:Goto 3
:Lbl Y
:Y+1→Y
:Goto 2
:Lbl N
:Goto 2
:Lbl D
:C-1->C
:ClrHome
:Disp "CLASS YES"
:Disp Y
:Disp "NO"
:Disp N
:Disp "STUDENTS"
:Disp C
:Y+L→L
:N+D→D
:Pause
:Goto1
```

TI-82

```
PROGRAM:SURVEY
:0→N
:0→Y
:0→S
:Lbl 2
:S+1→S
:Lbl 3
:ClrHome
:Disp "HAVE YOU LIVED A"
:Disp "MILLION SECONDS"
:Disp "PRESS ENTER TO"
:Disp "SEE CHOICES"
:Pause
:Menu("LIVED MILL.SEC", "YES", I,
   "NO", J, "CLASS COMPLETE", L)
:Lbl I
:Y+1→Y
:Goto 2
:Lbl J
:N+1→N
:Goto 2
:Lbl L
:ClrHome
:Output(1,1,Y)
:Output(2,1,"PEOPLE SAID YES")
:Output(4,1,N)
:Output(5,1,"SAID NO")
:Output(7,1,S-1)
:Output(7,4,"WERE SURVEYED")
:Stop
```

Unit Pricing

Change the last step to Stop for the TI-82.

TI-81and TI-82

```
Prgm:UNIT
:Disp "PRICE?"
:Fix 2
:Input A
:Disp "HOW MANY ITEMS?"
:Input B
:A/B→C
:Disp C
:Disp "NEXT PRICE?"
:Input D
:Disp "HOW MANY ITEMS?"
:Input E
:D/E→F
:If C < F
:Goto 2
:If C>F
:Goto 3
:Lbl 2
:Disp "BETTER BUY"
:Disp C
:Lbl 3
:Disp "BETTER BUY"
:Disp F
:End
```

Tessellations

This program will experiment with the tessellations of the rectangle. Try modifying the program by changing each expression $x + 1$ in the subroutine to $x + 2$ or $x + 3$. To create the subroutine, for {x} press [2nd] [0]; for {y} press [2nd] [·]. To access ClrDraw, press X(2nd) [DRAW] 1 [ENTER]; to access ClrStat, press [2nd] [STAT] [▶] [▶] 2 [ENTER]; to access xyLine, press [2nd] [STAT] [▶] 3 [ENTER].

TI-81

```
:ClrDraw
:0→Xmin
:15→Xmax
:0→Ymin
:10→Ymax
:1→Y
:Lbl B
:1→X
:Lbl A
:Prgm 2
:If X<15
:Goto A
:Y+1→Y
:If Y<10
:Goto B
:End
```

```
Prgm2:Square          These modifications will draw triangles.
:ClrDraw
:ClrStat
:X→{x}(1)
:Y→{y}(1)
:X+1→{x}(2)
:Y→{y}(2)
:X+1→{x}(3)           X+5→{x}(3)
:Y+1→{y}(3)
:X→{x}(4)
:Y+1→{y}(4)           Y→{y}(4)
:X→{x}(5)             Delete this line
:Y→{y}(5)             Delete this line
:xyLine
:X+1→X
:End
```

Star Fractal

This program explores a star fractal pattern. On the TI-82, turn the axes off with the command AxesOff.

TI-81 and TI-82

```
:π->G
:-.4→Xmin
:1.5→Xmax
:-.8→Ymin
:.7→Ymax
:5→P
:4→V
:144→A
:.35→R
:AG/180→A
:0→K
:0→L
:0→X
:0→Y
:0→N
:Disp N
:Lbl 1
:N→M
:N*A→B
:0→F
:Lbl 2
:If M/V≠Int M/V
:Goto 3
:If F≥P-1
:Goto 3
:F+1→F
:Int M/V→M
:Goto 2
:Lbl 3
:X+R^(P-F-1)cosB→X
:Y+R^(P-F-1)sinB→Y
:Line(K,L,X,Y)
:X→K
:Y→L
:N+1→N
:If N≤320
:Goto 1
:Pause
:ClrDraw
```